Lecture Notes in Mathematics

Edited by A. Dold and B. Eckmann

Subseries: Instituto de Matemática Pura e Aplicada, Rio de Janeiro
Adviser: C. Camacho

1195

J. Lucas M. Barbosa
A. Gervasio Colares

Minimal Surfaces in $\mathbb{R}^3$

Springer-Verlag
Berlin Heidelberg New York London Paris Tokyo

Authors

J. Lucas M. Barbosa
A. Gervasio Colares
Universidade Federal do Ceará, Departamento de Matemática
Campus do Pici, 60.000 Fortaleza-Ceará, Brasil

The final drawings from computer graphics and the other pictures were produced by Manfredo Perdigão do Carmo Jr.

This book is being published in a parallel edition by the Instituto de Matemática Pura e Aplicada, Rio de Janeiro as volume 40 of the series "Monografias de Matemática".

Mathematics Subject Classification (1980): 53C42

ISBN 3-540-16491-X Springer-Verlag Berlin Heidelberg New York
ISBN 0-387-16491-X Springer-Verlag New York Berlin Heidelberg

Printed in Germany

Printing and binding: Beltz Offsetdruck, Hemsbach/Bergstr.
2146/3140-543210

To S.-S. Chern

PREFACE

(Translated from the Brazilian edition)

This work constitutes the first step of a project whose purpose is to present examples of minimal surfaces with a minimum of theory. It has been written as Notes for a short course to be given in the "V Escola de Geometria Diferencial" at the "Universidade de São Paulo", from July 30th to August 3rd of 1984.

The Notes were prepared with a twofold purpose. First, to be useful to graduate students as an introduction to the study of minimal surfaces in R^3, and second, to constitute a catalogue of examples of such surfaces. The subject is quite wide and already has an excellent literature; thus, we did not have the intention of being exhaustive. On the contrary, we decided to orient the Notes towards complete minimal surfaces.

The necessary theory for the construction of the examples has been developed in a systematic form, trying to make the text as self-contained as possible. However, some of the more technical results were only described without proofs in order to avoid that the text would become overly long.

The first two chapters contain the classical results in the theory of minimal surfaces and present the examples of the catenoid, the helicoid and the surfaces of Scherk, Henneberg and Enneper. The Weierstrass Representation Formulas, which constitute the fundamental tool for the construction of examples, are derived there.

The remaining chapters contain the main results and examples obtained in the last thirty years about the theory of complete minimal surfaces. In particular, Chapter IV describes, with details, examples due to Jorge-Meeks, Meeks, Klotz-Sario, Costa and Chen-Gackstatter. The example of a complete minimal immersion of a torus minus 3 points into R^3, discovered by Costa, is presented from two points of view: using elliptic functions (as did by Costa himself) and by using the theory of hyperelliptic surfaces.

We regret we did not have the time to include in these Notes R. Schoen's results [1].

Part of the topics presented here has been discussed in a Seminar at the Department of Mathematics of "Universidade Federal do Ceará". We are grateful to it for support as well as to our colleagues and students in the Seminar, for the enthusiasm they have transmitted us, thereby creating the necessary atmosphere for the realization of this work. In particular, we want to thank Joaquim Rodrigues Feitosa for the expositions on Costa's examples.

The idea of writing these Notes was suggested to us by Professor Manfredo Perdigão do Carmo, who also participated in the plan of the work, and to whom we express our greatest aknowledgements. To our colleague José de Anchieta Delgado, we specially thank for valuable discussions and for having pointed out to us a short proof of a theorem of Jorge-Xavier.

We want to thank the Organizing Committee of the "V Escola de Geometria Diferencial" for the opportunity offered to us for presenting this work.

João Lucas Marques Barbosa[1]

Antonio Gervasio Colares[1]

---

1. The authors are professors at the Department of Mathematics of the "Universidade Federal do Ceará", in Brazil.

PREFACE TO THE ENGLISH LANGUAGE EDITION

After these notes appeared in Portuguese last year, several people asked us to publish an English version of it. Colleagues at IMPA were very enthusiastic about this idea and proposed its publication in the IMPA subseries of the Lecture Notes of Mathematics.

We did the translation and used the opportunity to make corrections and to include some additional material, such as M.E. de Oliveira's examples of nonorientable minimal immersions in R^3.

We also included several pictures of minimal surfaces. They were obtained by using a computer to produce the graphics from which the final drawings were made. The only exception is a picture of a Costa's minimal torus sent to us by David Hoffman.

We want to thank Manfredo Perdigão do Carmo for critical reading of the last version of the manuscript, Jonas Miranda Gomes for the time spent to devising the computer graphics from the equations and David Hoffman for permitting us to publish the picture of Costa's surface. Particular thanks are due to IMPA for computer facilities.

J. Lucas M. Barbosa

A. Gervasio Colares

TABLE OF CONTENTS

CHAPTER IV

RECENT EXAMPLES OF COMPLETE MINIMAL SURFACES

CHAPTER V

NONEXISTENCE OF CERTAIN MINIMAL SURFACES

CHAPTER I

A PRELIMINARY DISCUSSION

1. Introduction

It is generally admited that the investigations about minimal surfaces started with Lagrange [1] in 1760. He considered surfaces in R^3 that were graphics of C^2-differentiable functions $z = f(x,y)$. For such surfaces the area element is given by

$$(1.1) \qquad dM = (1 + f_x^2 + f_y^2)^{1/2} \, dx \wedge dy.$$

He studied the problem of determining a surface of this kind with the least possible area among all surfaces that assume given values on the boundary of an open set U of the plane (with compact closure and smooth boundary).

If $z = f(x,y)$ represents a solution for this problem, we consider a 1-parameter family of functions $z_t(x,y) = f(x,y) + t\eta(x,y)$, where η is a C^2-function that vanishes on the boundary of U, and we define

$$(1.2) \qquad A(t) = \int_{\bar{U}} (1 + (z_t)_x^2 + (z_t)_y^2)^{1/2} \, dxdy.$$

It follows that

$$A(t) = \int_{\bar{U}} \{(1+f_x^2+f_y^2) + 2t(f_x\eta_x+f_y\eta_y) + t^2(\eta_x^2+\eta_y^2)\}^{1/2} \, dxdy.$$

Set $p = f_x$, $q = f_y$ and $w = (1+p^2+q^2)^{1/2}$. Derivation with respect to t of the above equation gives

$$A'(0) = \int_{\bar{U}} 2(\frac{p}{w}\eta_x + \frac{q}{w}\eta_y)\,dM.$$

Integrating by parts and observing that $\eta|_{\partial\bar{U}} = 0$, we obtain

(1.3) $$A'(0) = -2\int_{\bar{U}} [\frac{\partial}{\partial x}(\frac{p}{w}) + \frac{\partial}{\partial y}(\frac{q}{w})]^{1/2}\,\eta\,dM.$$

Since $z = f(x,y)$ represents a solution for the problem, then $A(0)$ is a minimum for the function $A(t)$ and hence $A'(0) = 0$. This occurs for any function η chosen under the only restriction that η vanishes on the boundary of U. It follows that

$$\frac{\partial}{\partial x}(\frac{p}{w}) + \frac{\partial}{\partial y}(\frac{q}{w}) = 0.$$

By computing the indicated derivatives we obtain

(1.4) $$f_{xx}(1+f_y^2) - 2f_xf_yf_{xy} + f_{yy}(1+f_x^2) = 0.$$

This equation furnishes the necessary condition for one to solve the problem proposed by Lagrange. The solutions of the above equation were called "minimal surfaces", and they are given by real analytic functions.

Lagrange observed that a linear function (whose graphic is a plane) is clearly a solution for (1.4) and conjectured the existence of solutions containing any given curve given as a graphic along the boundary of U.

It was only in 1776 that Meusnier [1] gave a geometrical interpretation for (1.4) as meaning that

(1.5) $$H = \frac{k_1+k_2}{2} = 0,$$

where k_1 and k_2 stand for the principal curvatures introduced earlier by Euler. Furthermore, Meusnier also tried to find solutions for (1.4) endowed with special properties. For example, he determined the solutions of (1.4) whose level curves were straight lines. He did this as follows.

First he observed that when a curve is given implicitly by the equation $f(x,y) = c$, its curvature can be computed by

$$k = (-f_{xx}f_y^2 + 2f_xf_yf_{xy} - f_{yy}f_x^2)/|\text{grad } f|^3 . \tag{1.6}$$

Thus, one may rewrite equation (1.4) as

$$f_{xx} + f_{yy} = k|\text{grad } f|^3 . \tag{1.7}$$

If the level curves of f are straight lines, then $k \equiv 0$, and f is a harmonic function; that is, f satisfies the equation

$$\Delta f = \frac{\partial^2 f}{\partial x^2} + \frac{\partial^2 f}{\partial y^2} = 0 . \tag{1.8}$$

The only solutions for this equations whose level curves are straight lines are given by

$$f(x,y) = A \operatorname{arctg} \frac{y-y_o}{x-x_o} + B , \tag{1.9}$$

where A, B, x_o and y_o are constants. It is easily checked that the graphic of such functions is either a plane or a piece of a <u>helicoid</u> given by

$$\begin{cases} x - x_o = u \cos v \\ y - y_o = u \operatorname{sen} v \\ z - B = Av . \end{cases} \tag{1.10}$$

Meusnier also found the catenoid as the only minimal surface of revolution in R^3. (See next section.)

In 1835 Scherk [1] discovered another example of minimal surface, by solving the equation (1.4) for functions of the type $f(x,y) = g(x) + h(y)$. In this case, equation (1.4) reduces to

$$(1+h'^2(y))g''(x) + (1+g'^2(x))h''(y) = 0,$$

which is equivalent to

$$(1.11) \qquad -\frac{g''(x)}{1+g'^2(x)} = \frac{h''(y)}{1+h'^2(y)} .$$

Since x and y are independent variables, each side of this equation is constant. If a is this constant value, one obtains

$$(1.12) \qquad g(x) = \frac{1}{a}\log\cos ax, \qquad h(x) = -\frac{1}{a}\log\cos ay,$$

and hence $f(x,y) = \frac{1}{a}\log(\cos ax / \cos ay)$. The graphic of f is known as Scherk's minimal surface.

Scherk also tried, unsuccessfully, to determine all ruled minimal surfaces; that is, those minimal surfaces which contain a straight line through each one of its points. This problem was finally solved by Catalan [1] in 1842, who proved that the helicoid is the only ruled minimal surface in R^3.

The first general solution for the minimal surface equation was given by Weierstrass [1] in 1866, which allowed the construction of examples of minimal surfaces starting from the choice of two holomorphic functions. It is a direct consequence of this construction that minimal surfaces have real analytic coordinate functions.

In the next two sections we reobtain the classical examples of the catenoid and helicoid.

2. The Catenoid

The catenoid is a surface of revolution M in R^3 obtained by rotating the curve

$$\alpha(x) = (x, a \cosh (\frac{x}{a} + b)),$$

$x \in R$, around the x-axis.

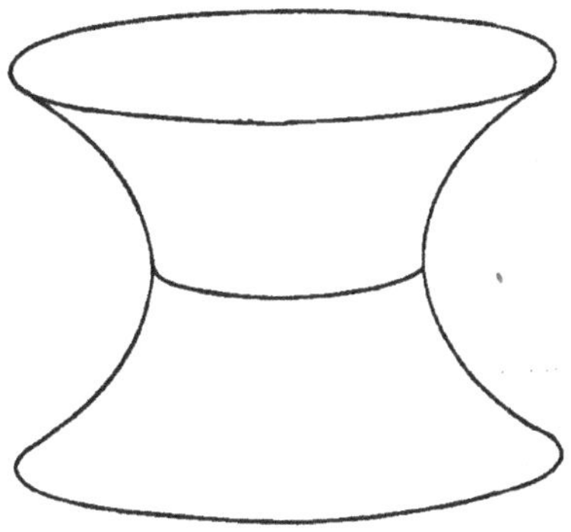

Fig. 1

Such a surface is minimal and complete. Its Gaussian curvature is $K = -1/a^2 \cosh^2 (\frac{x}{a} + b)$ and its total curvature, $\int_M KdM$, is -4π.

(2.1) THEOREM. <u>Any minimal surface of revolution in</u> R^3 <u>is, up to a rigid motion, part of a catenoid or part of a plane</u>.

<u>Proof</u>: By a rigid motion we may assume that the surface in R^3 is such that its rotation axis coincides with the x-axis. The surface will then be generated by a curve $\alpha(t) = (x(t), y(t), 0)$. If the function $x(t)$ is constant, then the surface will be a piece of a plane orthogonal to the x-axis. Otherwise, there exists a point t_o such that $x' \neq 0$ in a neighborhood of t_o. We may then represent α by

$$(x,\ y(x),\ 0)$$

in a neighborhood of the point $\alpha(t_o)$. The part of the surface obtained by rotating this piece of curve can be parametrized by

$$X(x,v) = (x,\ y(x)\cos v,\ y(x)\sin v).$$

It is a simple computation to show that $H = 0$ is then equivalent to

$$-yy'' + 1 + y'^2 = 0.$$

This equation can be integrated once by using the transformation $\frac{dy}{dx} = p$, from which it results that $\frac{d^2y}{dx^2} = \frac{dp}{dy}\frac{dy}{dx} = p\frac{dp}{dy}$. Substitution of this in the above equation yields

$$-y\,p\,\frac{dp}{dy} + 1 + p^2 = 0,$$

which can be easily integrated to give that

$$y = a\sqrt{1+p^2}.$$

A second integration now yields

$$\operatorname{arc\,cosh}\left(\frac{y}{a}\right) = \frac{x}{a} + b.$$

Therefore,

$$y = a\cosh\left(\frac{x}{a} + b\right).$$

Since minimal surfaces are real analytic, so is α. It follows that the curve α must coincide everywhere with the graphic of the above function $y(x)$. Hence the theorem is proved.

Some extensions of the above result can be found in D.Blair [1], J.L.M. Barbosa and M.P. do Carmo [1], H. Mori [1] and M.P. do Carmo and M. Dajczer [1].

3. The Helicoid

The helicoid can be described by the mapping $x: R^2 \to R^3$ given by

$$x(u,v) = (u \cos av, u \sin av, bv),$$

where a and b are nonzero constants. Geometrically, the helicoid is generated by a helicoidal motion of R^3 acting on a straight line parallel to the rotation plane of the motion.

The helicoid is a complete minimal surface. Its Gaussian curvature is $K = -b^2/(b^2+a^2u^2)^2$ and its total curvature is not finite.

The helicoid is also an example of a ruled surface; that is, of a surface described geometrically by a straight line sliding smoothly along a curve. (For a precise definition see M.P. do Carmo [1], pp. 188-189).

(3.1) THEOREM. Any ruled minimal surface of R^3 is, up to a rigid motion, part of a helicoid or part of a plane.

Proof: If $M \subset R^3$ is a ruled surface, then M can be parametrized locally by

(3.2) $$x(s,t) = \alpha(s) + t\beta(s),$$

where $\alpha(s)$ is a curve perpendicular to the straight lines of M and $\beta(s)$ describes a unit length vector field along α pointing in the direction of the straight line through $\alpha(s)$. We may assume that s represents the arc length of α and that α and β are analytic curves. A unit length normal vector field to x is

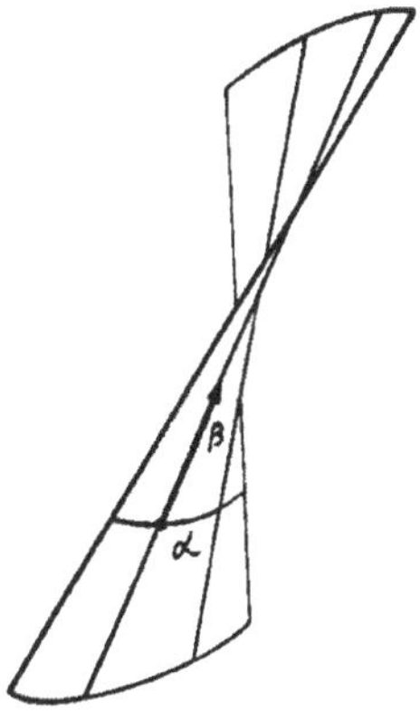

Fig. 2

given by

$$N = (\alpha' \times \beta + t\beta' \times \beta)/\sqrt{E},$$

where $E = 1 + 2t\langle \alpha', \beta' \rangle + t^2 |\beta'|^2$. It is easy to verify that M is a minimal surface if and only if

$$(3.3) \qquad \langle \alpha' \times \beta, \alpha'' \rangle + t\langle \beta' \times \beta, \alpha'' \rangle + t\langle \alpha' \times \beta, \beta'' \rangle + t^2 \langle \beta' \times \beta, \beta'' \rangle = 0.$$

Observing that the left hand side of this equation is a polynomial on the variable t, one obtains

$$(3.4) \qquad \begin{aligned} &(a) \quad \langle \alpha' \times \beta, \alpha'' \rangle = 0, \\ &(b) \quad \langle \beta' \times \beta, \alpha'' \rangle + \langle \alpha' \times \beta, \beta'' \rangle = 0 \quad \text{and} \\ &(c) \quad \langle \beta' \times \beta, \beta'' \rangle = 0. \end{aligned}$$

From equation (a) it follows that α'' must belong to the plane generated by α' and β. But, since α is parametrized by arc length, then α and α' are perpendicular. Hence,

(3.5) α'' <u>is parallel to</u> β.

It follows that $\langle \beta' \times \beta, \alpha'' \rangle = 0$ and so, equation (b) of (3.4) becomes simply

$$(b') \quad \langle \alpha' \times \beta, \beta'' \rangle = 0.$$

From equation (b') and (c) one concludes that

(3.6) β'' <u>belongs, simultaneously, to the planes generated by</u> α' <u>and</u> β, <u>and by</u> β' <u>and</u> β.

The intersection of these two planes contains, at least, the subspace generated by the vector β. If there exists a point where β'' is not parallel to β then, in a neighborhood of this point, these two planes coincide and α' is parallel to β'. Since α and β are real analytic functions, this occurs everywhere. Hence, we have that $(\beta \times \alpha')' = \beta' \times \alpha' + \beta \times \alpha'' = 0$. Thus, the plane generated by β and α' is constant. Therefore, α is a plane curve and the surface described by x is a plane.

On the other hand, if β'' is parallel to β everywhere and α' and β' are not parallel at one point, then this occurs in a neighborhood of this point. In this case we claim that

(3.7) <u>the curvature and the torsion of</u> α <u>are constant</u>.

In fact, since $k = \langle \alpha'', \beta \rangle$, we have

$$\pm \frac{dk}{ds} = \langle \alpha'', \beta \rangle' = -\langle \alpha', \beta' \rangle' = -\langle \alpha'', \beta' \rangle - \langle \alpha', \beta'' \rangle = 0.$$

It is easy to see that $\pm\tau = \langle \alpha' \times \beta', \beta \rangle$ and that

$$\pm \frac{d\tau}{ds} = \langle \alpha' \times \beta', \beta \rangle' = \langle \alpha'' \times \beta', \beta \rangle + \langle \alpha' \times \beta'', \beta \rangle + \langle \alpha' \times \beta', \beta' \rangle = 0.$$

Hence, k and τ are constants. It follows that, up to a rigid motion of R^3, α can be parametrized by

$$\alpha(s) = (A \cos as,\ A \sin as,\ bs),$$

where $A^2a^2 + b^2 = 1$. Since β is parallel to α'', $\beta(s) = \pm (\cos as, \sin as, 0)$. If we take $u = A \pm t$ and $v = s$, then (3.2) becomes

$$x(u,v) = (u \cos as,\ u \sin as,\ bv).$$

Therefore, M is a piece of a helicoid. This proves the theorem.

Another proof can be found in M.P. do Carmo [1] (p.197).

For an extension of the above result the reader is refered to J.L.M. Barbosa, M. Dajczer and L.P.M. Jorge [1].

CHAPTER II

THE WEIERSTRASS REPRESENTATION AND THE CLASSICAL EXAMPLES

1. The Weierstrass representation

Let $U \subset R^2$ be a simply connected open set and $\psi: U \to R^3$ be an immersion of class C^k $(k \geq 2)$ or real analytic. The mapping ψ describes a parametric surface in R^3. If

(1.1) $$|\psi_u| = |\psi_v| \quad \text{and} \quad \langle \psi_u, \psi_v \rangle = 0,$$

then ψ is a conformal mapping (i.e., it preserves angles) which induces in U the metric

(1.2) $$ds^2 = \lambda^2(du^2+dv^2),$$

where $\lambda = |\psi_u| = |\psi_v|$. We then say that (u,v) are isothermal parameters for the surface described by ψ.

(1.3) THEOREM (Existence of isothermal parameters). *Let U be a simply connected open set and let $\psi: U \to R^3$ be an immersion of class C^k, $k \geq 2$ (or real analytic). Then, there exists a diffeomorphism $\varphi: U \to U$ of class C^k (or real analytic) such that $\tilde{\psi} = \psi \circ \varphi$ is a conformal mapping.*

A proof os this theorem may be found in Spivak [1], vol. 4.

Let M be a surface (i.e., a 2-dimensional manifold of class

C^k). Suppose M is connected and orientable and let $x: M \to R^3$ be an immersion of class C^k. By the above theorem each point $p \in M$ has a neighborhood in which isothermal parameters (u,v) are defined. The metric induced on M by x will be represented, locally, in terms of such parameters, by

$$ds^2 = \lambda^2 |dz|^2 , \tag{1.4}$$

where $z = u+iv$. Clearly, a change of coordinates of such parameters is a conformal mapping.

Since M is orientable, we can restrict ourselves to a family of isothermal parameters whose changes of coordinates preserve the orientation of the plane. In terms of the variable $z = x+iy$, this means that such changes of coordinates are holomorphic. A surface M together with such a family of isothermal parameters is called a Riemann surface.

We can extend the notion of holomorphic mapping to such surfaces as follows: if M and $\tilde{M}$ are Riemann surfaces, we say that $f: M \to \tilde{M}$ is holomorphic when every of its representation, in terms of local isothermal parameters (in M and $\tilde{M}$), is a holomorphic function.

In a Riemann surface we consider, locally, the operators

$$\frac{\partial}{\partial z} = \frac{1}{2} \left(\frac{\partial}{\partial u} - i \frac{\partial}{\partial v}\right) \quad \text{and} \quad \frac{\partial}{\partial \bar{z}} = \frac{1}{2} \left(\frac{\partial}{\partial u} + i \frac{\partial}{\partial v}\right) . \tag{1.5}$$

The definition of these operators is such that, if $f: M \to \mathbb{C}$ is a complex valued differentiable function, then

$$df = \frac{\partial f}{\partial u} du + \frac{\partial f}{\partial v} dv = \frac{\partial f}{\partial z} dz + \frac{\partial f}{\partial \bar{z}} d\bar{z} .$$

(1.6) The function f is holomorphic if and only if $\frac{\partial f}{\partial \bar{z}} = 0$; if $\frac{\partial f}{\partial z} = 0$ we say that f is anti-holomorphic.

The expression of many entities in the study of surfaces are simplified when considered in the context of Riemann surfaces. For example, the Laplace operator becomes

(1.7) $$\Delta = \frac{1}{\lambda^2}\left(\frac{\partial}{\partial u^2} + \frac{\partial}{\partial v^2}\right) = \frac{4}{\lambda^2}\frac{\partial}{\partial z}\frac{\partial}{\partial \bar{z}} .$$

Another example is the Gaussian curvature of M which is then given by

(1.8) $$K = -\Delta \log \lambda .$$

For an immersion $x = (x_1, x_2, x_3): M \to R^3$, we define Δx as the vector valued function $(\Delta x_1, \Delta x_2, \Delta x_3)$. Then,

(1.9) $$\Delta x = 2HN ,$$

where H is the mean curvature of the immersion and $N: M \to S^2(1)$ is its Gauss mapping.

To prove this equation, we first observe that Δx is normal to M. In fact, it follows from (1.1) that

$$\langle x_u, x_u\rangle = \langle x_v, x_v\rangle \quad \text{and} \quad \langle x_u, x_v\rangle = 0 .$$

By derivation, we obtain

$$\langle x_{uu}, x_u\rangle = \langle x_{uv}, x_v\rangle \quad \text{and} \quad \langle x_{vu}, x_v\rangle + \langle x_u, x_{vv}\rangle = 0.$$

Hence,

$$\langle x_{uu}+x_{vv}, x_u\rangle = \langle x_{uv}, x_v\rangle - \langle x_{vu}, x_v\rangle = 0 .$$

Similarly, one shows that

$$\langle x_{uu}+x_{vv}, x_v\rangle = 0,$$

and, from this, it comes that Δx is normal to M.

We recall that if N is the Gauss mapping of the immersion x, and $N_u = a_{11}x_u + a_{12}x_v$ and $N_v = a_{21}x_u + a_{22}x_v$, the mean curvature is given by

$$H = -\frac{1}{2}(a_{11}+a_{22}).$$

Now, by using (1.7) we obtain

$$\lambda^2\langle \Delta x, N\rangle = \langle x_{uu}+x_{vv}, N\rangle = -\langle x_u, N_u\rangle - \langle x_v, N_v\rangle =$$

$$= -a_{11}|x_u|^2 - a_{22}|x_v|^2 = -(a_{11}+a_{22})\lambda^2 = 2H\lambda^2,$$

thus proving (1.9).

The proposition below is an immediate corollary of (1.9).

(1.10) PROPOSITION. <u>A mapping</u> $x: M \to R^3$ <u>is a minimal immersion if and only if</u> x <u>is harmonic</u>.

Define $\phi = \frac{\partial x}{\partial z}$. From (1.7) we get $\frac{\partial \phi}{\partial \bar{z}} = \frac{\partial}{\bar{z}}\frac{\partial x}{\partial z} = \frac{\lambda^2}{4}\Delta x$. Hence,

(1.11) ϕ <u>is holomorphic if and only if</u> x <u>is harmonic</u>.

Note that ϕ is a function defined locally on M with values in $\mathbb{C}^3$. In fact, its image is contained in a quadric Q of $\mathbb{C}^3$ given by the equation

(1.12) $$z_1^2 + z_2^2 + z_3^2 = 0, \qquad (z_1, z_2, z_3) \in \mathbb{C}^3 .$$

To see this we observe that if $\phi = (\phi_1, \phi_2, \phi_3)$, then

$$\phi_k = \frac{1}{2}\left(\frac{\partial x_k}{\partial u} - i\frac{\partial x_k}{\partial v}\right).$$

Hence,

$$\sum_{k=1}^{3} \phi_k^2 = \frac{1}{4}\{\sum_{k=1}^{3}(\frac{\partial x_k}{\partial u})^2 - \sum_{k=1}^{3}(\frac{\partial x_k}{\partial v})^2 - 2i\sum_{k=1}^{3}\frac{\partial x_k}{\partial u}\frac{\partial x_k}{\partial v}\} =$$

$$= \frac{1}{4}\{|x_u|^2 - |x_v|^2 - 2i\langle x_u, x_v\rangle\} = 0.$$

Therefore,

$$(1.13) \qquad \sum_{k=1}^{3} \phi_k^2 = 0.$$

Similarly, we obtain

$$(1.14) \qquad |\phi|^2 = \sum_{k=1}^{3} |\phi_k|^2 = 2\lambda^2 .$$

Thus

$$|\phi| > 0.$$

Now, observe that we have a mapping ϕ defined, in terms of isothermal parameters, in some neighborhood of each point of M. If $z = x+iy$ and $w = r+is$ are isothermal parameters around some point in M, then the change of coordinates $w = w(z)$ is holomorphic with $\frac{\partial w}{\partial z} \neq 0$. It follows that $\tilde{\phi} = \frac{\partial x}{\partial z}$ is related to ϕ by

$$\phi = \frac{\partial x}{\partial z} = \frac{\partial x}{\partial w}\frac{\partial w}{\partial z} = \frac{\partial w}{\partial z}\tilde{\phi} .$$

Thus, if we consider the vector valued differential forms $\alpha = \phi dz$ and $\tilde{\alpha} = \tilde{\phi} dw$, we have

$$\alpha = \phi dz = \tilde{\phi}\frac{\partial w}{\partial z} dz = \tilde{\phi} dw = \tilde{\alpha} .$$

This means that we have a vector valued differential form α globally defined on M, whose local expression is $\alpha = (\alpha_1, \alpha_2, \alpha_3)$, with

$$\alpha_k = \phi_k dz , \qquad 1 \le k \le 3.$$

This, together with (1.10) and (1.11) prove the following

(1.15) PROPOSITION. *Let* $x: M \to R^3$ *be an immersion. Then,* $\alpha = \phi dz$ *is a vector valued holomorphic form on* M *if and only if* x *is a minimal immersion. Moreover,*

$$x = \mathrm{Re}\left(\int^z \alpha\right),$$

where the integral is taken along any path from a fixed point to z *on* M.

When the real part of the integral of α along any closed path is zero, we say that α has no *real periods*. The nonexistence of real periods for α is easily seen to be equivalent to $\mathrm{Re}(\int^z \alpha)$ be independent of the path on M.

(1.16) THEOREM (Weierstrass representation). *Let* $\alpha_1, \alpha_2, \alpha_3$ *be holomorphic differentials on* M *such that*

(a) $\Sigma\, \alpha_k^2 = 0$ (*i.e., locally* $\alpha_k = \phi_k dz$ *and* $\Sigma\, \phi_k^2 = 0$);

(b) $\Sigma |\alpha_k|^2 > 0$ *and*

(c) *each* α_k *has no real periods on* M.

Then, the mapping $x: M \to R^3$ *defined by* $x = (x_1, x_2, x_3)$, *with* $x_k = \mathrm{Re}(\int^z \alpha_k)$, *is a minimal immersion.*

The conditions (c) of the theorem is necessary to guarantee that $\mathrm{Re}(\int_{p_0}^z \alpha_k)$ depends only on the final point z. Thus, each x_k is well defined independently of the path from p_0 to z. It is clear that $\phi = \frac{\partial x}{\partial z}$ is holomorphic and so x is harmonic. Hence, x is minimal. The condition (b) guarantees that x is an immersion.

It is possible to give a simple description of all solutions of the equation $\alpha_1^2 + \alpha_2^2 + \alpha_3^2 = 0$ on M. For this, we suppose that $\alpha_1 \neq i\alpha_2$. (If $\alpha_1 = i\alpha_2$, then $\alpha_3 = 0$ and the resulting minimal surface is a plane.) Now we define a holomorphic form ω and a meromorphic function g by

$$(1.17)\qquad \omega = \alpha_1 - i\alpha_2 , \qquad g = \frac{\alpha_3}{\alpha_1 - i\alpha_2} .$$

Locally, if $\alpha_k = \phi_k dz$, then $\omega = f\,dz$, where f is a holomorphic function and we have

$$(1.18)\qquad f = \phi_1 - i\phi_2 , \qquad g = \frac{\phi_3}{\phi_1 - i\phi_2} .$$

In terms of g and ω, the forms α_1, α_2 and α_3 can be reobtained as

$$(1.19)\qquad \alpha_1 = \frac{1}{2}(1-g^2)\omega , \qquad \alpha_2 = \frac{i}{2}(1+g^2)\omega , \qquad \alpha_3 = g\omega .$$

Therefore, the minimal immersion x of the above theorem is given by

$$(1.19')\qquad \begin{aligned} x_1 &= \mathrm{Re}\left(\int^z \alpha_1\right) = \mathrm{Re}\int^z \frac{1}{2}(1-g^2)\omega \\ x_2 &= \mathrm{Re}\left(\int^z \alpha_2\right) = \mathrm{Re}\int^z \frac{i}{2}(1+g^2)\omega \\ x_3 &= \mathrm{Re}\left(\int^z \alpha_3\right) = \mathrm{Re}\int^z g\omega . \end{aligned}$$

If z_0 is a point where g has a pole of order m, then, from (1.19) it is clear that ω must have a zero of order exactly $2m$ at z_0, in order to have condition (b) satisfied and each α_k holomorphic.

(1.20) Conversely, suppose we have defined on M a meromorphic function g and a holomorphic form ω, whose zeroes coincide with the poles of g, in such way that each zero of order m of ω corresponds to a pole of order $2m$ of g. Then, the forms α_1, α_2 and α_3, defined as above, are holomorphic on M, satisfy $\alpha_1^2 + \alpha_2^2 + \alpha_3^2 = 0$ and

$$(1.21) \qquad |\phi|^2|dz|^2 = \Sigma|\alpha_k|^2 = \frac{1}{2}(1+|g|^2)^2 |\omega|^2 > 0.$$

Furthermore, if such forms α_k $(1 \le k \le 3)$, do not have real periods, then we may apply (1.19') to obtain a minimal immersion $x\colon M \to R^3$.

The equations (1.19') are called Weierstrass representation formulas for minimal surfaces in R^3. This representation enables us to describe a great number of examples of minimal surfaces. The expression of the metric obtained "a posteriori" of such a representation is

$$(1.22) \qquad ds^2 = \frac{1}{2}|f|^2 (1+|g|^2)^2 |dz|^2 .$$

The meromorphic function $g\colon M \to \mathbb{C} \cup \{\infty\}$ which appears in the Weierstrass representation of a minimal immersion $x\colon M \to R^3$ has an important geometrical meaning. To see this, let us obtain an expression for the Gauss mapping $N\colon M \to S^2(1)$ in terms of the Weierstrass representation for x. Locally, at each point of M, $\alpha_k = \phi_k dz$ define the functions ϕ_k and, from (1.19'), we then get

$$x_u \times x_v = -(\mathrm{Re}\,\phi_1, \mathrm{Re}\,\phi_2, \mathrm{Re}\,\phi_3) \wedge (\mathrm{Im}\,\phi_1, \mathrm{Im}\,\phi_2, \mathrm{Im}\,\phi_3) =$$
$$= (\mathrm{Im}\,\phi_2\bar\phi_3, \mathrm{Im}\,\phi_3\bar\phi_1, \mathrm{Im}\,\phi_1\bar\phi_2) =$$
$$= \frac{1}{4}|f|^2 (1+|g|^2)(2\,\mathrm{Re}\,g, 2\,\mathrm{Im}\,g, g^2-1) .$$

It follows that

(1.23) $$N = \left(\frac{2\ \mathrm{Re}\ g}{1+|g|^2}, \frac{2\ \mathrm{Im}\ g}{1+|g|^2}, \frac{|g|^2-1}{|g|^2+1}\right) .$$

If $\pi: S^2(1)-\{(0,0,1)\} \to R^2$ is the stereographic projection, then $\pi \circ N = (\mathrm{Re}\ g, \mathrm{Im}\ g)$ at every point of M, except at the poles of g. If we identify R^2 with the complex plane $\mathbb{C}$ and extend π to a mapping $\tilde{\pi}: S^2(1) \to \mathbb{C} \cup \{\infty\}$ with $\pi((0,0,1)) = \infty$, then

(1.24) $$\tilde{\pi} \circ N = g .$$

This means that the mapping g can be identified with the Gauss mapping of x. A direct computation using (1.8), (1.14) and (1.21), yields the following value for the Gaussian curvature of M:

(1.25) $$K = - \left[\frac{4|g'|}{|f|(1+|g|^2)^2}\right]^2 .$$

Since g' is holomorphic, we have the following corollary of (1.25):

(1.26) <u>Either</u> $K = 0$ <u>or its zeroes are isolated</u>.

In the next sections we will use the Weierstrass representation to reobtain the Helicoid, the Catenoid and Scherk's surface. We also reobtain the surfaces of Enneper and Henneberg.

2. The Helicoid

Take $M = \mathbb{C}$, $g(z) = -ie^z$ and $\omega = e^{-z}\,dz$. Observe that neither g has poles nor ω has zeroes in $\mathbb{C}$. By (1.19)

$$\alpha_1 = \frac{1}{2}(1-g^2)\omega = \cosh(z)dz,$$

(2.1)
$$\alpha_2 = \frac{i}{2}(1+g^2)\omega = -i\ \sinh(z)dz,$$

$$\alpha_3 = g\omega = -i\ dz.$$

Since $\cosh(z)$, $\sinh(z)$ and multiplication by a constant are holomorphic functions in $\mathbb{C}$, we have that $\int_\gamma \alpha_k = 0$, for every closed path γ in $\mathbb{C}$ and $k = 1,2,3$. That is, the forms α_k do not have periods. From (1.19') we obtain

$$x_1 = \mathrm{Re}\int_0^z \cosh(z)dz = \mathrm{Re}(\sinh(z)) = \cos(v)\ \sinh(u),$$

(2.2)
$$x_2 = \mathrm{Re}\int_0^z -i\ \sinh(z)dz = \mathrm{Re}(-i\ \cosh(z)+i) = \sin(v)\ \sinh(u),$$

$$x_3 = \mathrm{Re}\int_0^z -idz = \mathrm{Re}(-iz) = v.$$

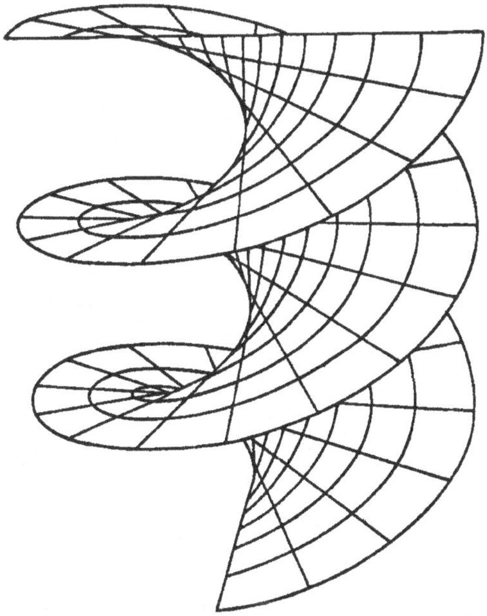

Fig. 3 - The Helicoid

Thus, $x(u,v) = (\cos(v)\sinh(u), \sin(v)\sinh(u), v)$ describes a minimal immersion. Making $\sinh(u) = t$, the immersion $x(t,v)$ is exactly the helicoid described in the Section 3 of Chapter I.

3. The Catenoid

Take $M = \mathbb{C}$, $g(z) = -e^{z}$ and $\omega = -e^{-z}\,dz$. Observe that g has no poles and ω has no zeroes in $\mathbb{C}$. From (1.19) we have

$$\begin{aligned} \alpha_1 &= \sinh(z)dz \;, \\ \alpha_2 &= -i\cosh(z)dz \;, \\ \alpha_3 &= dz \;. \end{aligned} \tag{3.1}$$

Since $\cosh(z)$, $\sinh(z)$ and multiplication by a constant are holomorphic functions in $\mathbb{C}$, then $\int_\gamma \alpha_k = 0$, for each closed path γ in $\mathbb{C}$, $k = 1,2,3$. Hence, the forms α_k have no periods. From (1.19') we obtain

$$\begin{aligned} x_1 &= \mathrm{Re}\int_0^z \sinh(z)dz = \mathrm{Re}(\cosh(z)-1) = \cos(v)\cos(u) - 1, \\ x_2 &= \mathrm{Re}\int_0^z -i\cosh(z)dz = \mathrm{Re}(-i\sinh(z)) = \sin(v)\cosh(u) \;, \\ x_3 &= \mathrm{Re}\int_0^z dz = \mathrm{Re}(z) = u \;. \end{aligned} \tag{3.2}$$

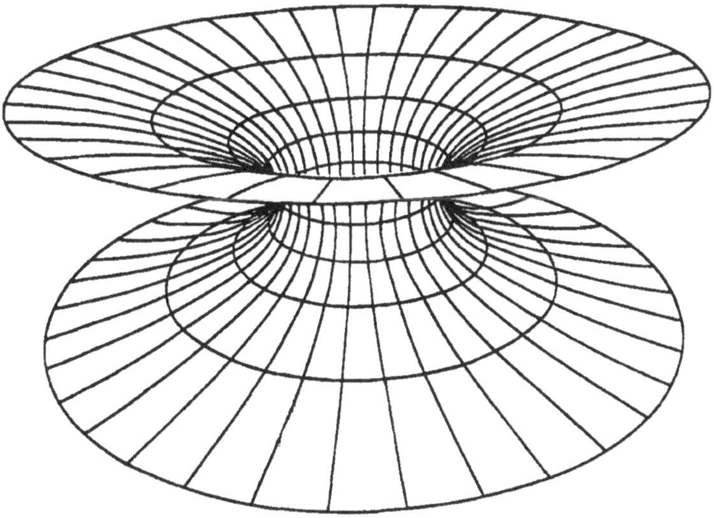

Fig. 4 - The Catenoid

Thus, $x(u,v) = (\cos(v)\cosh(u), \sin(v)\cosh(u), u) - (1,0,0)$. This is, up to a translation, the parametrization of the catenoid described in Chapter I, Section 2. Such a parametrization wraps the plane $\mathbb{C}$ around the catenoid infinitely many times.

Another way of obtaining the catenoid is the following: take $M = \mathbb{C}-\{0\}$, $g(z) = z$ and $\omega = dz/z^2$. Then,

$$\alpha_1 = \frac{1}{2}\left(\frac{1}{z^2} - 1\right)dz ,$$

(3.3)
$$\alpha_2 = \frac{i}{2}\left(\frac{1}{z^2} + 1\right)dz ,$$

$$\alpha_3 = \frac{1}{z}\,dz .$$

The forms α_1 and α_2 do not have periods and α_3 has only a purely imaginary period. By using (1.19') we obtain

$$\text{(3.4)}\qquad \begin{aligned} x_1 &= -\frac{u}{2}\left(1+\frac{1}{u^2+v^2}\right)+1 , \\ x_2 &= -\frac{v}{2}\left(1+\frac{1}{u^2+v^2}\right) , \\ x_3 &= \frac{1}{2}\log\left(u^2+v^2\right) . \end{aligned}$$

These equations describe the catenoid, up to a translation. To see this, set

$$\rho = \frac{1}{2}\log(u^2+v^2) \quad \text{and} \quad \theta = \left(\operatorname{arctg}\frac{v}{u}\right) - \pi .$$

4. Scherk's surface

Consider the unit disk $D = \{z \in \mathbb{C};\ |z| < 1\}$. Take $M = D$, $g(z) = z$ and $\omega = 4dz/(1-z^4)$. From (1.19) we obtain

$$\text{(4.1)}\qquad \begin{aligned} \alpha_1 &= \frac{2dz}{1+z^2} = \left(\frac{i}{z+i} - \frac{i}{z-i}\right)dz , \\ \alpha_2 &= \frac{2idz}{1-z^2} = \left(\frac{i}{z+1} - \frac{i}{z-1}\right)dz , \\ \alpha_3 &= \frac{4z}{1-z^4}\,dz = \left(\frac{2z}{z^2+1} - \frac{2z}{z^2-1}\right)dz . \end{aligned}$$

Clearly, α_1, α_2 and α_3 have no periods in D. From (1.19') we get

$$\text{(4.2)}\qquad \begin{aligned} x_1 &= \operatorname{Re}\left(i\log\frac{z+i}{z-i}\right) = -\arg\left(\frac{z+i}{z-i}\right) , \\ x_2 &= \operatorname{Re}\left(i\log\frac{z+1}{z-1}\right) = -\arg\left(\frac{z+1}{z-1}\right) , \\ x_3 &= \operatorname{Re}\left(\log\frac{z^2+1}{z^2-1}\right) = \log\left|\frac{z^2+1}{z^2-1}\right| . \end{aligned}$$

It is easy to see that

$$\frac{z+i}{z-i} = \frac{|z|^2-1}{|z-i|^2} + i\,\frac{z+\bar{z}}{|z-i|^2}$$

and

$$\frac{z+1}{z-1} = \frac{|z|^2-1}{|z-1|^2} + \frac{z-\bar{z}}{|z-1|^2} .$$

Since $|z|^2-1 < 1$ in D, we have that $-\frac{3\pi}{2} \leq x_j \leq -\frac{\pi}{2}$, $j = 1,2$, $z = x_1+ix_2$. It is also straightforward from the above expressions that

$$\cos x_1 = \frac{|z|^2-1}{z^2+1} \quad \text{and} \quad \cos x_2 = \frac{|z|^2-1}{z^2-1},$$

which yield

$$x_3 = \log\left(\frac{\cos x_2}{\cos x_1}\right) ,$$

where (x_1,x_2) is restricted to $(-\frac{3\pi}{2},-\frac{\pi}{2})(-\frac{3\pi}{2}, -\frac{\pi}{2})$. Thus the

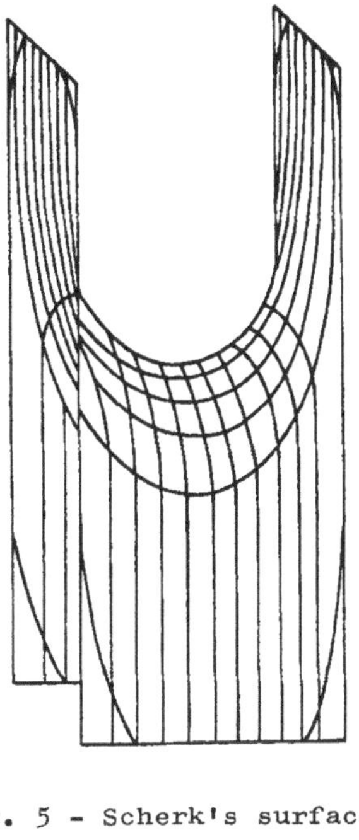

Fig. 5 - Scherk's surface

immersion $x: D \to R^3$ describes a piece of the Scherk's minimal surface mentioned in the first chapter. To obtain the whole surface we consider g and ω defined in $M = \mathbb{C} - \{1,-1,i,-i\}$. Here the resulting forms α_1 and α_2 have real periods. Let then $\pi: \tilde{M} \to M$ be the universal covering of M and define

$$\tilde{x}_k = \mathrm{Re} \int^z \pi^*\alpha_k , \qquad k = 1,2,3.$$

Since $\tilde{M}$ is simply connected, the forms $\pi^*\alpha_k$ have no periods and so, the functions $\tilde{x}_k: \tilde{M} \to R$ are well defined, $k = 1,2,3$. The image $\tilde{x}(\tilde{M})$ can be obtained from the previous functions x_k if we now allow $x_1 = -\arg\left(\frac{z+i}{z-i}\right)$ and $x_2 = -\arg\left(\frac{z+1}{z-1}\right)$ to assume all possible values under the only restriction that

$$\frac{\cos x_2}{\cos x_1} > 0.$$

This is equivalent to consider the entire graphic of the real function

$$x_3 = \log\,(\cos x_2/\cos x_1).$$

5. Enneper's surface

The simplest choice that one can make for M, g and ω is to take $M = \mathbb{C}$, $g(z) = z$ and $\omega = dz$. It results a minimal immersions $x: \mathbb{C} \to R^3$ given by

$$(5.1) \qquad x(u,v) = \frac{1}{2}\left(u - \frac{u^3}{3} + uv^2,\ \ -v + \frac{v^3}{3} - u^2v,\ \ u^2 - v^2\right),$$

which describes the Enneper's surface.

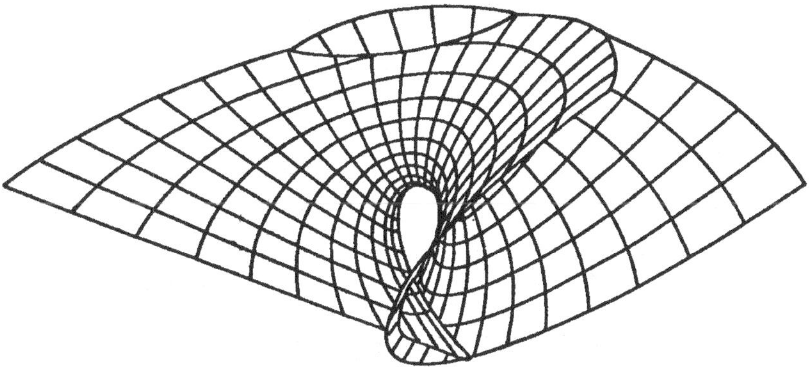

Fig. 6 - Enneper's surface

This is a complete minimal surface. Its Gaussian curvature is

(5.2) $$K = -\frac{16}{(1+|z|^2)^4} ,$$

where $z = u+iv$.

6. Henneberg's surface

Take $M = \mathbb{C}-\{0\}$, $g(z) = z$ and $\omega = 2(1 - \frac{1}{z^4})dz$. We obtain

(6.1)
$$\alpha_1 = (-\frac{1}{z^4} + \frac{1}{z^2} + 1 - z^2)dz ,$$
$$\alpha_2 = i(-\frac{1}{z^4} - \frac{1}{z^2} + 1 - z^2)dz ,$$
$$\alpha_3 = 2(z - \frac{1}{z^3})dz .$$

Observe that α_1, α_2 and α_3 have no periods in M. We then obtain

$$\int_1^z \alpha_1 = \frac{(1-z^2)^3}{3z^3} ,$$

$$\int_1^z \alpha_2 = \frac{i(1+z^2)^3}{3z^3} - \frac{8i}{3} ,$$

$$\int_1^z \alpha_3 = \frac{(z^2-1)^2}{z^2} .$$

Now, from (1.19') we get

$$(6.2) \qquad \begin{aligned} x_1 &= \mathrm{Re}\ (\bar{z}-|z|^2 z)^3/3|z|^6 , \\ x_2 &= -\mathrm{Im}\ (\bar{z}+|z|^2 z)^3/3|z|^6 , \\ x_3 &= \mathrm{Re}\ (z|z|^2-\bar{z})^2/|z|^4 ; \end{aligned}$$

that is,

$$(6.3) \qquad \begin{aligned} x_1 &= \{u^3(1-u^2-v^2)^3-3uv^2(1-u^2-v^2)(1+u^2+v^2)^2\}/3(u^2+v^2)^3 , \\ x_2 &= \{3u^2v(1+u^2+v^2)^2(1-u^2-v^2) - v^3(1-u^2-v^2)^3\}/3(u^2+v^2)^3 , \\ x_3 &= \{(1-u^2-v^2)^2u^2 - (1+u^2+v^2)v^2\}/(u^2+v^2)^2 . \end{aligned}$$

Now, let $\varphi(z) = (1-z^2)/z$ and $\psi(z) = (1+z^2)/z$. It is easy to verify that

$$(6.4) \qquad \varphi(-\tfrac{1}{\bar{z}}) = \overline{\varphi(z)} \quad \text{and} \quad \psi = (-\tfrac{1}{\bar{z}}) = \overline{\psi(z)} .$$

Since $x_1 = \frac{1}{3}\mathrm{Re}(\varphi(z)^3)$, $x_2 = \frac{1}{3}\mathrm{Re}\ i(\psi(z)^3)$ and $x_3 = \mathrm{Re}(\varphi(z)^2)$, we have that $x_k(-1/\bar{z}) = x_k(z)$, $k = 1,2,3$. If we identify M with the unit sphere minus two points through the stereographic projection, then z and $-1/\bar{z}$ correspond to antipodal points on the sphere.

From (6.4) we conclude that $x = (x_1, x_2, x_3)$ can be looked upon as a mapping from the projective plane into R^3. Therefore, $x(M)$ is a Möbius strip in R^3.

Unfortunately, x is not regular at every point; in fact, we have $\Sigma|\alpha_k|^2 = 0$ at the points ± 1 and $\pm i$, which are the only singular points of x. Since these represent two pairs of antipodal points, we then have that x, considered as a mapping on the projective plane, is singular at exactly two points.

Thus, x restricted to $\mathbb{C} - \{0,1,-1,i,-i\}$ represents a minimal immersion which is not complete and whose image is a Möbius strip minus two points. Since $g(z) = z$, its Gauss mapping covers each point of $S^2(1)$ just once, with exception of six points. Therefore, its total curvature is -4π.

CHAPTER III

COMPLETE MINIMAL SURFACES WITH FINITE TOTAL CURVATURE

1. Complete minimal surfaces

The examples of minimal surfaces considered so far (the catenoid, the helicoid, Scherk's surface and Enneper's surface), with the only exception of Henneberg's surface, are all complete in the induced metric.

The seach for examples of complete minimal surfaces began with geometrically simple examples. In 1915 S. Bernstein proved the following

(1.1) THEOREM (Bernstein). *If* $f: R^2 \to R$ *is a* C^2*-differentiable function whose graphic is a minimal surface, then* f *is linear*.

This result is a beautiful and non-trivial example of a global theorem in partial differential equations. Much work was devoted in trying to generalize it. The reader may obtain further information about the proof and generalizations of this theorem in Osserman [1], Bombieri-Giorgi-Giusti [1] and do Carmo-Peng [1].

Even for the case of surfaces in R^3, this theorem was improved, and stronger results were obtained. For this, it was necessary to express the above result in a slightly different way, as follows:

(1.2) THEOREM. *If* M *is a complete minimal surface in* R^3 *whose normals form an acute angle with a fixed direction, then* M *is a plane.*

This formulation of Bernstein's Theorem was generalized by R. Osserman [4] and later by F. Xavier [1].

(1.3) THEOREM (F. Xavier). *Let* M *be a complete minimal surface in* R^3 *and* $N: M \to S^2(1)$ *be its Gauss mapping. If* $N(M)$ *omits seven or more points, then* M *is a plane.*

Osserman's result, as well as Xavier's result, makes use of the Weierstrass representation and of a deep theorem in Complex Analysis, known as Koebe uniformization theorem. The formulation of this theorem that will be useful to us is the following:

(1.4) THEOREM (Uniformization). *Let* M *be a Riemann surface endowed with a complete metric* ds^2. *Let* Δ *represent any one of the following surfaces: the unit sphere, or the complex plane* $\mathbb{C}$, *or the unit disk* D. *Then, there exists a locally invertible conformal mapping* F *from* Δ *onto* M.

Of course, if $x: M \to R^3$ is a complete minimal immersion, then, making use of the above theorem, we may consider the function $x \circ F: \Delta \to R^3$, which will still be a complete minimal immersion in the induced metric.

Since minimal immersions in R^3 can not be compact (because $K \leq 0$), Δ can never be a sphere. Hence, we may always restrict ourselves to the cases $\Delta = \mathbb{C}$ or $\Delta = D$. Thus, for $x \circ F$, we will have a global Weierstrass representation on $\mathbb{C}$ or D.

One application of these observations is the proof of a generalization of Theorem (1.2), which can be restated as follows.

(1.5) THEOREM (Osserman [4]). *If* $x: M \to R^3$ *is a complete minimal immersion, then the image of the Gauss mapping* N *of* x *is dense in* $S^2(1)$, *unless* $x(M)$ *is a plane.*

Proof: By the above observations we may assume that $x: \Delta \to R^3$, where $\Delta = \mathbb{C}$ or $\Delta = D$. If $N(\Delta)$ is not dense in $S^2(1)$, there exist $P \in S^2(1)$ and ε, $1 > \varepsilon > 0$, such that

$$\langle N,P \rangle \le 1-\varepsilon .$$

By changing coordinates in R^3, we may assume that $P = (0,0,1)$. Consider the Weierstrass representation of the immersion. By using the expression for N given in (II 1.23) we conclude that $|g(z)| \le \le A < \infty$. If $\Delta = \mathbb{C}$, by Liouville's theorem, g is constant. Hence, the Gauss mapping N is constant and $x(M)$ describes a plane. If $\Delta = D$, we can only conclude that g has no poles, and so f has no zeroes. If α is any curve in D starting at the origin and going to the boundary of D, we have

$$\text{length}(\alpha) = \int_\alpha ds = \frac{1}{2}\int_\alpha |f|(1+|g|^2)|dz| < \frac{1+A^2}{2}\int_\alpha |f||dz| .$$

We will show that there exists α of finite length, thus contradicting the hypothesis of completeness of M. For this, consider the function

$$x = \int_0^z f(\xi)d\xi .$$

Since $f \neq 0$ in D, $w: D \to \mathbb{C}$ is locally invertible. Let $z = G(w)$ be a local inverse function for w in a small disk around $w = 0$. Let R be the radius of the largest disk where G can be defined. Clearly, $R < \infty$ since the image of G lies in the disk $|z| < 1$. Thus, there exists a point w_0 with $|w_0| = R$ such that G can not be extended to a neighborhood of w_o. Set $\ell = \{tw_o;\ 0 \le t < 1\}$ and $\alpha = G(\ell)$. The curve α so defined is divergent. In fact, if it were not, there would exist a sequence $\{t_n\}$ converging to 1 such that the corresponding sequence $\{z_n\}$ along α would converge to a point z_o in D. By continuity, $G(w_o) = z_o$. But then, since the function w is invertible at z_o, G would be extendable to a neighborhood of w_o, a contradiction. Therefore, α is divergent. On the other hand,

$$\int_\alpha |f(z)||dz| = \int_0^1 |f(z)||\frac{dz}{dt}|dt = \int_0^1 |\frac{dw}{dt}|dt = \int_\ell |dw| = R < \infty .$$

Thus, lenth(α) $< \infty$ and so, with the induced metric, D is not complete, a contradiction. This completes the proof of Osserman's result.

The proof of Xavier's Theorem uses the results below.

(1.6) PROPOSITION. _Let_ $g\colon D \to \mathbb{C}-\{0,a\}$, $(a\neq 0)$, _be a holomorphic function_. _Set_ $\alpha = 1-1/k$, _where_ k _is a natural number_. _Then_

$$\int_D \frac{|g'(z)|^p dxdy}{[|g(z)|^\alpha + |g(z)|^{2-\alpha}]^p} < \infty ,$$

for any p, $0 < p < 1$.

(1.7) THEOREM (Yau [1]). _Let_ M _be a complete surface of infinite volume_. _If_ $u\colon M \to \mathbb{C}$ _is a non-negative function satisfying_ $\Delta \log u = 0$ _almost everywhere, then_ $\int_M u^p\, dM = \infty$, _for any positive_ p.

Proposition (1.6) depends on classical results about complex functions on the disk. Its proof can be found in Xavier [1]. Theorem (1.7) is a particular case of a more general result proved by Yau [1] (p. 661).

The proof of Xavier's theorem starts as the proof of Theorem (1.5). We may assume $x: \Delta \to R^3$, where $\Delta = \mathbb{C}$ or $\Delta = D$. When $\Delta = \mathbb{C}$ the function g of the Weierstrass representation will be a meromorphic function $g: \mathbb{C} \to \mathbb{C}$ which, by the (small) Picard theorem, can omit at most 2 points of $\mathbb{C}$, unless it is a constant. Since g and N are essentially the same mapping, then N omits at most 3 points (of $S^2(1)$), unless it is constant.

When $\Delta = D$ Picard's theorem can not be used. If $p_1,\dots,p_k,p_{k+1} = \infty$ are the points omited by $g: \mathbb{C} \to \mathbb{C} \cup \{\infty\}$, we define

$$h = g'/f^q|(g-p_1)(g-p_2)\dots(g-p_k)|^\alpha ,$$

where f is the other holomorphic function associated with the Weierstrass representation of x. The function h is also holomorphic and, in particular, $\Delta \log|h| = 0$, almost everywhere on D (g' can vanish at points of a discrete set). Thus, by Yau's theorem we must have

$$(1.8) \qquad \int_D |h|^p dM = \frac{1}{4}\int_D |h|^p(1+|g|^2)^2|f|^2 \, dudv = \infty ,$$

for any choice of $p > 0$. Now, let us estimate the value of $\int_D |h|^p dM$ in a different way. We choose ϵ, $1 > \epsilon > 0$, sufficiently small, so that the closed sets $D_j = \{z \in D;\ |g(z)-p_i| \le \epsilon\}$, $1 \le j \le k-1$, are disjoint. Define $\hat{D} = D - \bigcup_{j=1}^{k-1} D_j$. Then,

$$(1.9) \qquad \int_D |h|^p dM = \int_{\hat{D}} |h|^p dM + \sum_{j=1}^{k-1} \int_{D_j} |h|^p dM .$$

The function $|g-p_j|^{\alpha p}(1+|g|^2)^2/|f|^{pq-2}\ |g-p_1|^{\alpha p}\dots|g-p_k|^{\alpha p}$ is continuous in D_j and since D_j is compact, it is bounded by a constant C. It follows that

(1.10) $$\int_{D_j} |h|^p \, dM \le C \int_{D_j} \frac{|g'|^p}{|g-p_j|^{\alpha p}} \, dudv \le$$

$$\le C \int_D 2^p \frac{|g'|^p \, dudv}{(|g-p_j|^{\alpha}+|g-p_j|^{2-\alpha})^p} \, .$$

In $\hat{D}$, the function $|g-p_k|^{(k-1)p\alpha-4} (1+|g|^2)^2/|f|^{pq-2}|g-p_1|^{p\alpha} \ldots$ $\ldots |g-p_k|^{p\alpha}$ is continuous and will be bounded if $q = 2/p$ and $(k-1)p\alpha \ge 4$. If, furthermore, it is possible to choose α and p in such way that $kp\alpha = 5$, then

(1.11) $$\int_{\hat{D}} |h|^p \, dM \le C \int_{\hat{D}} \frac{|g'|^p dudv}{|g-p_k|^{kp\alpha-4}} \le$$

$$\le \int_{\hat{D}} \frac{2^p|g'|^p \, dudv}{(|g-p_k|^{\alpha}+|g-p_k|^{2-\alpha})^p} \, .$$

From (1.9), (1.10), (1.11) and making use of (1.6), we obtain

(1.12) $$\int_D |h|^p \, dM < \infty \, ,$$

for any p satisfying $0 < p < 1$, $p \ge 4/(k-1)\alpha$ and $kp\alpha = 5$. This contradicts (1.8).

To conclude the proof of the theorem let us examine more carefully the conditions about p which have led us to this contradiction. Since $kp\alpha = 5$ and $0 < p < 1$, then $\alpha > 5/k$. Since $\alpha < 1$, then $k > 5$. Hence, the first possible choice for k is $k = 6$. For this value of k and for $1 > \alpha > 5/6$ it is possible to choose p satisfying to $1 > p > 4/(k-1)^{\alpha}$ and $p\alpha = 5/k = 5/6$. For example, $\alpha = 6/7$ and $p = 35/36$. Thus, we obtain a contradiction for $k+1 \ge 7$. Therefore the theorem is proved.

We ougth to point out to the reader the existence of complete minimal surfaces in R^3 whose Gauss mapping omits a set of k points, for any $0 \leq k \leq 4$. The Gauss mapping of Enneper's surface omits one single point, that of the catenoid omits two points, and that of Scherk's surface omits four points. To these examples we may add the examples below.

(1.12) C.C. Chen's surface. This is obtained by considering $M = \mathbb{C}$, $g(z) = z + \frac{1}{z}$ and $\omega = z^2 dz$. This surface is complete and $N(M) = S^2(1)$.

(1.13) K. Voss surfaces. Take $M = \mathbb{C} - \{p_1, \ldots, p_m\}$, $g(z) = z$ and $\omega = dz/(z-p_1)\ldots(z-p_m)$. This is a minimal surface whose Gauss mapping omits $m+1$ points. This surface is complete if $m \leq 3$. Observe that Scherk's surface is a particular case of a K. Voss surface.

It is a very interesting open question to determine if Xavier's theorem is the best possible. On this direction are known the following two results which will not be proved here.

(1.14) THEOREM (Osserman [5]). *If M is a complete minimal surface in R^3 with finite total curvature that is not a plane, then $N(M)$ omits at most three points.*

(1.15) THEOREM (Gackstätter [2]). *If M is a complete abelian minimal surface in R^3 that is not a plane, then $N(M)$ omits at most four points.*

A minimal surface is called *abelian* when it can be constructed by using a compact Riemann surface $\hat{M}$, a meromorphic function $g: \hat{M} \to \mathbb{C} \cup \{\infty\}$ and a meromorphic form ω as follows. One considers the forms $\alpha_1, \alpha_2, \alpha_3$, defined in (II 1.19) and the open set

$R = \{p \in \hat{M};\ 0 < \sum_{j=1}^{3} |\alpha_j(p)|^2 < \infty\}$. Let M be the universal covering space of R and $\pi: M \to R$ be the corresponding covering mapping. One then defines $x: M \to R^3$ by $x = (x_1, x_2, x_3)$, where

$$x_k = \mathrm{Re}\int^z \pi^*\alpha_k, \qquad 1 \le k \le 3.$$

It is straightforward to show that x describes a complete minimal immersion in R^3.

Theorem (1.14) and (1.15) are also proved in Chen and Simões [1]. An open question is to know whether it is possible to prove these results by using similar ideas as the ones in the proof of Xavier's theorem. It is not known whether Theorem (1.14) is the best possible. To answer this question one should either exhibit an example of complete minimal surface with finite total curvature whose Gauss mapping omits exactly 3 points, or improve that theorem for "two points".

A deep question in the study of complete minimal surfaces in R^3 is the following:

(Calabi) Does there exist a bounded complete minimal surfaces in R^3 ?

In 1980, Jorge and Xavier [1] exhibited a non-trivial example of a complete minimal surface lying between two parallel planes in R^3. This example is constructed in the next chapter. They also proved in [2], the following result:

(1.16) THEOREM. There are no bounded complete minimal surfaces in R^3 with bounded Gaussian curvature.

A proof for this theorem (indicated by J. Anchieta Delgado) is obtained by using the following result:

(1.17) THEOREM (Omori [1]). *Let* M *be a complete surface with Gaussian curvature bounded from below.* *If* $f: M \to R$ *is a function bounded from above then, for each* $p \in M$ *and each* $\varepsilon > 0$, *there exists* $q \in M$ *such that*

i) $f(q) \geq f(p)$

ii) $|\mathrm{grad}\, f_q| < \varepsilon$ *and*

iii) $\mathrm{Hess}\, f_q(V,V) \leq \varepsilon |V|^2$, *for each* $V \in T_qM$.

This theorem generalizes the well known result that any real differentiable function defined in a compact surface attains a maximum. We will apply it to obtain (1.16).

Suppose there exists a bounded minimal immersion $x: M \to R^3$ such that M, endowed with the induced metric, is complete and has $|K| < c$. Define a function $f: M \to R$ by $f(p) = \frac{1}{2}|x(p)-x(p_o)|^2$, where p_o is a fixed point of M. A simple computation gives us

$$\mathrm{Hess}\, f_p(V,V) = |V|^2 + \langle II(V,V), x(p)-x_o\rangle \ , \tag{1.18}$$

where II is the second fundamental form of the immersion x, $x_o = x(p_o)$ and $V \in T_pM$. By Omori's theorem there exist points $p_1,\dots,p_m,\dots$ such that: $f(p_m) \geq f(p)$, $|\mathrm{grad}\, f_{p_m}| < 1/m$ and $\mathrm{Hess}\, f_{p_m}(V,V) \leq \frac{1}{m}|V|^2$, for each V in $T_{p_m}M$. It follows that

$$\langle II(V,V), x(p_m)-x_o\rangle \leq (\tfrac{1}{m} - 1)|V|^2 < 0$$

and hence the mean curvature H of M satisfy

$$H(p_m)\langle N(p_m), x(p_m)-x_o\rangle < 0.$$

Since such an inequality can not occur when $H = 0$, the proof is complete.

2. Complete minimal surfaces with finite total curvature

The study of complete minimal surfaces with finite total curvature begins with the following result:

(2.1) THEOREM (Osserman [5]). *Let* M *be a complete surface in* R^3 *whose Gaussian curvature satisfy*

a) $K \leq 0$

b) $\int_M |K| dM < \infty$.

Then, there exists a compact surface $\hat{M}$, *a finite number of points* $p_1, \ldots, p_k$ *of* $\hat{M}$ *and an isometry from* M *onto* $\hat{M} - \{p_1, \ldots, p_k\}$.

This is a deep result and its proof is not presented in these notes. For the proof, see Osserman [1].

Complete minimal surfaces of R^3 with finite total curvature satisfy the hypotheses of the above theorem. Furthermore, after identification of M and $\hat{M} - \{p_1, \ldots, p_k\}$, the Gauss mapping $N: M \to S^2(1)$, which is conformal, extends to a meromorphic function $\hat{N}: \hat{M} \to S^2(1)$. In fact, if any of the points p_j were an essential singularity of N (N is identified with the function g of the Weierstrass representation) then, by Picard's (great) theorem, N would assume all values of $S^2(1)$ infinitely many times, with at most two exceptions. But this would imply that the total curvature of M would be infinite, which is contrary to hypothesis (b). Thus, at each point p_j, N has at most a pole; hence, it can be extended as a meromorphic function to $\hat{M}$. We also have that the form ω extends as a meromorphic form to $\hat{M}$. Indeed, by changing coordinates in R^3, we may assume that g has no poles at the points $p_1, \ldots, p_k$. Since the metric of M is complete we have

$$\lim_{z \to p_j} |\omega| = \infty .$$

Thus, the only singularities of ω are poles and so ω is a meromorphic form on $\hat{M}$.

Meromorphic functions of a compact surface into $S^2(1)$ have the property of assuming each value the same (finite) number of times (counting multiplicity). As a consequence, we have the result below.

(2.2) PROPOSITION. Let M be a complete minimal surface in R^3 with finite total curvature. Then

$$\int_M |K| \, dM = -4m\pi ,$$

where m is a nonnegative integer.

The above value of the total curvature is subject to restrictions originated from the topology of M. Such a statement is justified by the following

(2.3) THEOREM (Osserman [5]). Let $\hat{M}$ be a compact surface and $x: \hat{M} - \{p_1, \ldots, p_k\} \to R^3$ be a complete minimal immersion with total curvature $-4m\pi$. Then

$$2m \geq 2k - \chi(\hat{M}),$$

where $\chi(\hat{M})$ represents the Euler characteristic of $\hat{M}$.

The proof of this theorem makes use of the following lemma, which deals with the forms α_i refered to in (II 1.16):

(2.4) LEMMA. Under the hypotheses of the previous theorem the form $\alpha = (\alpha_1, \alpha_2, \alpha_3)$ has a pole of order $m_j \geq 2$ at each p_j.

<u>Proof of the lemma:</u> Around p_j, α_i can be represented by

$$\alpha_i = \phi_i(z)dz$$

with $z = 0$ corresponding to p_j. At $z = 0$, each α_i has at most a pole of order m_{ij}. Since $\hat{M} - \{p_1,\ldots,p_k\}$ is complete in the induced metric and this metric is given by

$$ds^2 = \frac{1}{2}\sum_{i=1}^{3} |\phi_i(z)|^2 \, |dz|^2$$

in the neighborhood under consideration, then

$$\lim_{z\to 0} \sum_{i=1}^{3} |\phi_i(z)|^2 = \infty.$$

Thus, $m_j = \max\{m_{1j},m_{2j},m_{3j}\} \geq 1$. Observe that m_j is exactly the order of the pole of α at p_j.

Assume that $m_j = 1$. Then,

$$\alpha_i(z) = \frac{c_i}{z} + b_i + \ldots , \qquad i = 1,2,3.$$

Since $x_i = \mathrm{Re}\int \phi_i(z)dz$, it follows that $x_i - \mathrm{Re}(c_i \log z)$ is also well defined, and so, c_i must be real. Taking into account that $\sum_{i=1}^{3} \phi_i^2 = 0$, then, $\Sigma\, c_i^2 = 0$, and so each c_i is zero. But this is a contradiction, since $m_j \geq 1$. Therefore, the lemma is proved.

<u>Proof of the theorem:</u> Since the extension of g to $\hat{M}$ assumes each value of $S^2(1)$ exactly m times, counting multiplicities, then (after a change of coordinates of R^3) we can suppose that $g(p_j) \neq \infty$ and 0, $1 \leq j \leq k$. It follows from the previous lemma together with (II 1.19) that the form ω has a pole of order $m_j \geq 2$ at p_j and exactly $2m$ zeroes (counting multiplicities). By Riemann's relation we have

number of poles of ω - number of zeroes of $\omega = \chi(\hat{M})$

(cf. Ahlfors and Sario [1], V.27A). Thus, we obtain

(2.5) $$\chi(\hat{M}) = \sum_{j=1}^{r} m_j - 2m \geq 2k - 2m.$$

Hence,

$$2m \geq 2k - \chi(\hat{M}),$$

thus proving the theorem.

Let us observe that the Euler characteristic of M is given by

(2.6) $$\chi(M) = \chi(\hat{M}) - k.$$

Thus, we obtain the following

(2.7) COROLLARY (Chern-Osserman's inequality). Set $M = \hat{M}-\{p_1,\ldots,p_k\}$ as in the previous theorem and let $x\colon M \to R^3$ be a complete minimal immersion with finite total curvature. Then,

$$\int_M K\,dM \leq 2\pi(\chi(M)-k).$$

The proposition below is now a consequence of the Chern-Osserman inequality).

(2.8) PROPOSITION. A complete minimal surface in R^3 with finite total curvature -4π is the catenoid or Enneper's surface.

Proof: This is the case of $m=1$. This means that the function g is meromorphic of order 1, hence it transforms $\hat{M}$ conformally onto $S^2(1)$. Thus, $\chi(\hat{M}) = 2$. It follows from Theorem (2.3) that $k \leq 2$. Therefore, we may choose M as being $S^2(1)$ minus one or two points.

Since g is 1-1 and locally invertible, g is a diffeomorphism. We may then introduce conformal coordinates in M by using g^{-1}. This is equivalent to consider $\hat{M} = S^2(1) = \mathbb{C} \cup \{\infty\}$ and $g(z) = z$.

<u>1st case</u>: M is $S^2(1)$ minus one point. We may then assume that $M = \mathbb{C}$. The form ω is now given by $f(z)dz$, where f is a rational function without poles in $\mathbb{C}$. On the other hand, $f \neq 0$ on M (otherwise, the metric of M would be singular). Therefore, f must be constant, and the surface obtained from the Weierstrass representation is the Enneper's surface.

<u>2nd case</u>: M is $S^2(1)$ minus two points. We may then assume that $M = \mathbb{C}-\{a\}$. The form ω is given by $f(z)dz$, where f is a rational function with a pole at $z = a$ (if not, the metric of M would not be complete). On the other hand, $f \neq 0$ on M, otherwise the metric of M would be singular. Thus, $f(z) = c/(z-a)^n$, where we may assume that a is real. By Lemma (2.4), we must have $n \geq 2$. Since any path going to ∞ must have infinite length, $n=2$ or $n=3$. The case of $n=3$ is discarded by observing that the periods of α_i would be

$$\int_{|z-a|=\epsilon} \alpha_1 = -ic\pi, \qquad \int_{|z-a|=\epsilon} \alpha_2 = c\pi \qquad \int_{|z-a|=\epsilon} \alpha_3 = 0$$

and the only possibility for a choice of c would be $c = 0$, which is not possible. When $n = 2$, a similar computation of periods yields $-iac\pi$, πac and zero. Hence, we must have $a = 0$ and the surface obtained from the Weierstrass representation is the catenoid.

(2.9) An <u>end</u> of an immersed surface is a part of the surface which is homeomorphic to a topological disk punctured at its center such that every path on this disk which diverges to its center has infinite length.

Let $\hat{M}$ be a compact surface, $p_1,\dots,p_k$ a finite number of points in $\hat{M}$ and $x\colon \hat{M}-\{p_1,\dots,p_k\} \to R^3$ a complete minimal immersion. If $D \subset \hat{M}$ is a neighborhood containig p_j, then the image $x(D-\{p_j\})$ of $D-\{p_j\}$ by x is an end of $X(\hat{M}-\{p_1,\dots,p_k\})$ which is denoted by E_j. Thus, $x(\hat{M}-\{p_1,\dots,p_k\})$ is an immersed surface in R^3 with k ends.

The catenoid is an example of a complete minimal surface in R^3 with two ends. By letting the catenoid M intersect a sphere $S^2(r)$ centered at the origin and with radius r, the ends of the catenoid are then given by the two connected componentes E_1 and E_2 of $M - S^2(r)$ which lie in the exterior of $S^2(r)$. As we have seen the catenoid is described by a mapping $x\colon \hat{M}-\{p_1,p_2\} \to R^3$ where $\hat{M}$ is the unit sphere $S^2(1)$, and its Gauss mapping $g\colon S^2(1)-\{p_1,p_2\}\to S^2(1)$ extends conformally to $S^2(1)$. In this sense we say that the catenoid has a well defined normal vector at infinity on each end.

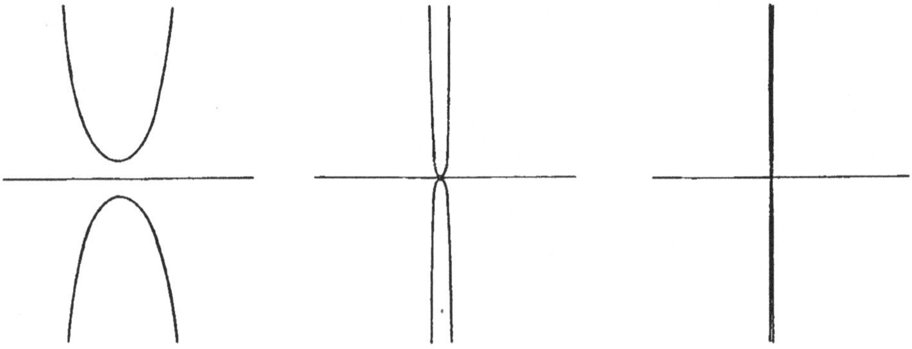

(a) (b) (c)

Fig. 7 - Profiles of the Catenoid as an observer moves to infinity

If Y_r represents the intersection of the catenoid with the sphere $S^2(r)$, then Y_r/r converges to a great circle in $S^2(1)$, as r goes to infinity. Such a fact leads to the conclusion that the catenoid viewed from infinity looks like two copies of a plane passing through the origin with opposite orientations. (See Fig.7(c)).

Inspired in these facts, Jorge and Meeks [1] proved the following

(2.10) THEOREM. Let M be a complete surface immersed in R^3, diffeomorphic to $\hat{M} - \{p_1,\dots,p_k\}$, where $\hat{M}$ is an orientable compact surface such that the Gauss mapping extends continuously to $\hat{M}$. If $Y_r = M \cap S^2(r)$, then Y_r/r consists of closed curves $\Gamma_1,\dots,\Gamma_k$ in $S^2(1)$ which converge C^1 to closed geodesics $\gamma_1,\dots,\gamma_k$ of $S^2(1)$, with multiplicities $I_1,\dots,I_k$, as r goes to infinity.

We will not present here the proof of this theorem; however, we will make some remarks about it.

First of all, observe that the hypothesis that M is homeomorphic to a compact surface minus k points means that M has k ends. Next, the condition about the Gauss mapping is equivalent to say that M has well defined normal vector at infinity on each end. Let E_j represent the end of M corresponding to p_j, that is, E_j is the image of a punctured neighborhood of p_j in $\hat{M}$. The idea of the theorem is that, as one "gets close" to p_j in $\hat{M}$, then the tangent plane to $\hat{M}$ at p_j "gets close" to the plane perpendicular to p_j passing through the origin. To make sense this "closeness" one considers the quotient Y_r/r and makes r grows to infinity. Finally, it is a consequence of this theorem that a complete minimal surface in R^3 with finite total curvature, viewed from infinity, looks like k planes passing through the origin. Each one of such planes corresponds to an end.

The above theorem allows the following geometrical interpretation of Chern-Osserman's inequality (2.7).

(2.11) THEOREM (Jorge-Meeks [1]). *Let* M *be a complete minimal surface immersed in* R^3 *with finite total curvature* $-4m\pi$ *and having* k *ends* (*i.e.* M *is diffeomorphic to* $\hat{M} - \{p_1, \ldots, p_k\}$, *where* $\hat{M}$ *is compact*). *Then*,

$$2m = \sum_{j=1}^{k} I_j - \chi(M) \geq k - \chi(M),$$

where I_j *is the multiplicity of the end* E_j *corresponding to* p_j. *Equality holds if and only if each end is embedded.*

Proof: From the previous theorem, for each end E_j of M, we have that

$$\Gamma_r^j = \frac{1}{r}(E_j \cap S^2(r))$$

is a closed curve immersed in $S^2(1)$ and

$$\lim_{r\to\infty} \Gamma_r^j = \gamma^j ,$$

where γ^j is a closed geodesic in $S^2(1)$ with multiplicity I_j and the convergence is C^1. Thus, $I_j = 1$ if and only if Γ_r^j is an embedded closed curve for r sufficiently large. Moreover, since the convergence of Γ_r^j to γ^j is C^1, as $r \to \infty$, the total curvature of Γ_r^j converges to the total curvature of γ^j, which is $2\pi I_j$.

Let B_r be the ball of R^3 of radius r and center at the origin; take $M_r = \frac{1}{r}(M \cap B_r)$. By the previous theorem, the plane of γ^j and the tangent space to E_j at infinity coincide. Since the convergence is C^1, as r goes to infinity the total geodesic curvature of Γ_r^j, as the boundary curve of M_r, converges to the total geodesic curvature $2\pi I_j$ of γ^j. That is,

$$(2.12) \qquad \lim_{r\to\infty} C(\Gamma_r^j) = 2\pi I_j ,$$

where $C(\Gamma_r^j)$ is the total geodesic curvature of Γ_r^j.

Denoting by $C(M_r)$ the total curvature of M_r and applying Gauss-Bonnet's formula to M_r we obtain

$$(2.13) \qquad C(M_r) + \sum_{j=1}^{k} C(\Gamma_r^j) = 2\pi\chi(M_r) = 2\pi\chi(M).$$

As $r \to \infty$, this simplifies to

$$(2.14) \qquad 2\pi\chi(M) = C(M) + 2\pi \sum_{j=1}^{k} I_j .$$

Therefore,

$$(2.15) \qquad C(M) = 2\pi(\chi(M) - \sum_{j=1}^{k} I_j).$$

Since $C(M) = -4m\pi$, one obtains

$$(2.16) \qquad 2m = \sum_{j=1}^{k} I_j - \chi(M) \geq k - \chi(M) ,$$

where we have used that $I_j \geq 1$. Finally, $\sum_{j=1}^{k} I_j = k$ if and only if $I_j = 1$, for every $j = 1,2,\ldots,k$; that is, the equality holds if and only if each end E_j is embedded, $j = 1,\ldots,k$. This proves the theorem.

(2.17) COROLLARY. The catenoid is the only embedded minimal annulus in R^3 with finite total curvature.

Proof: From (2.15), the total curvature of an embedded minimal annulus in R^3 with finite total curvature is -4π. In fact, we have $k = 2$ and $\chi(M) = 0$, hence $m = 1$. We know that if the total curvature is -4π, then M is the catenoid or Enneper's surface. Therefore, in this case, M is the catenoid, completing the proof.

An open question related to this subject is the following: Which are the complete embedded minimal surfaces in R^3 ?

Until quite recently, the plane, the helicoid and the catenoid were the only known examples. In his doctoral dissertation, C. Costa [1], using elliptic functions, exhibited an example of a complete minimal immersion of a torus minus three points in R^3 (cfr. IV, 7). Later, Hoffman and Meeks [1] proved that Costa's surface is one of a large family of examples of embedded minimal surfaces.

By using Jorge and Meeks ideas one can prove the following

(2.18) PROPOSITION. <u>A complete minimal surface immersed in R^3 with finite total curvature and only one end which is embedded is a plane.</u>

<u>Proof</u>: Suppose $M \subset R^3$ is an immersed surface satisfying the hypotheses of the proposition. By (2.10), the closed curves

$$\Gamma_r = \frac{1}{r}(M \cap S_r^2),$$

converge, with multiplicity, to a closed geodesic γ of $S^2(1)$, as $r \to \infty$. Since the end of M is embedded, for r sufficiently large, also Γ_r is embedded and the multiplicity is one. This implies that, for r sufficiently large, the end of M is a graphic over the plane π of γ. To prove this write

$$M = M_r \cup E_r,$$

where E_r is the end of M given by the part of M exterior to S_r^2 and M_r is the complement of E_r in M_r. Since the Gauss mapping of M has a limit as $r \to \infty$, E_r projects injectively over the plane π. On the other hand, for r sufficiently large, Γ_r projects orthogonally over a convex curve in the plane π of γ. The theorem below due to Radó can then be applied to guarantee that also M_r is a graphic over π.

(2.19) THEOREM (Radó [1]). If a Jordan curve $\Gamma \subset R^3$ admits an injective orthogonal projection over a convex curve in a plane $R^2 \subset R^3$, then there exists a unique compact minimal surface M having Γ as boundary, which is a graphic of a real function on R^2.(*)

It follows that M is a graphic in R^3 and since M is complete, by Bernstein's theorem, M is a plane, thus completing the proof.

We want to point out that Enneper's surface has finite total curvature (-4π) and only one end, but this does not contradicts the above proposition because the end is not embedded.

We should remark that the hypothesis that M is orientable has been implicitly assumed in all the above statements. When M is not orientable, we must pass to the oriented two-sheeted covering space M' of M. If $\pi: M' \to M$ is the covering mapping and $x: M \to R^3$ is a complete minimal immersion, then $x \circ \pi: M' \to R^3$ is again a complete minimal immersion. We then can define the total curvature $C(M)$ of M by

$$(2.20) \qquad C(M) = \frac{1}{2}\int_{M'} K'\,dM'.$$

Let the Gauss mapping of M be defined as a function which associates to each point of M its normal direction considered as a point in the projective plane $\mathbb{P}^2$. Then, it is clear that

$$C(M) = -2\pi m,$$

where m counts the number of times that the Gauss mapping of M

(*) A simple proof of Radó's theorem can be found in Lawson [1].

covers $\mathbb{P}^2$. In fact, m is also the number of times that the Gauss mapping of $x \circ \pi$ covers $S^2(1)$. Observe that the Proposition (2.8) implies the nonexistence of complete minimal immersions of nonorientable surfaces into R^3 with $C(M) = -2\pi$ (since the catenoid and Enneper's surface are orientable). A theorem of Meeks [4] also allows us to conclude the nonexistence of minimal immersions with $C(M) = -4\pi$. In the same paper he exhibits an example of a Möbius strip minimally immersed in R^3 with total curvature -6π; this will be described in the next chapter.

The basic result in the study of complete minimal immersions of nonorientable surfaces into R^3 is the theorem below whose proof is omitted.

(2.22) THEOREM (Meeks [4]). Let $\hat{M}$ be a nonorientable compact surface and $x: \hat{M} - \{p_1,\dots,p_k\} \to R^3$ a complete minimal immersion. If the total curvature $C(M)$ is finite, then

$$\frac{C(M)}{2\pi} \equiv \chi(\hat{M}) \mod 2.$$

By using this theorem and some results obtained before, we will prove the following

(2.23) THEOREM OF CLASSIFICATION (Osserman-Jorge-Meeks). Let M be a complete minimal surface immersed in R^3 with total curvature greater than -8π. Then, up to a projective transformation of R^3, M is the plane, the catenoid, Enneper's surface or Meeks minimal Möbius strip.

Proof: From Proposition (2.8), the catenoid and Enneper's minimal surface are the only orientable complete minimal surfaces immersed in R^3 with total curvature -2π. Consequently, there does not exist a nonorientable complete minimal surface immersed in R^3 with total curvature -2π.

We will show now that there does not exist a nonorientable complete minimal surface immersed in R^3 with total curvature -4π.

Set $M = \hat{M} - \{p_1,\dots,p_k\}$ and let $x\colon \hat{M} - \{p_1,\dots,p_k\} \to R^3$ be a complete minimal immersion of a nonorientable surface with total curvature -4π. By the above theorem, the Euler characteristic $\chi(\hat{M})$ is even. Since $\hat{M}$ is nonorientable we have $\chi(\hat{M}) \le 0$. By applying the formula of Jorge-Meeks (cfr. (2.11)) to the oriented two-sheeted covering $M' = \hat{M}' - \{p_1,\dots,p_{2k}\}$, of M, we obtain

$$2m \ge 2k - \chi(M') = 4k - \chi(\hat{M}'). \tag{2.24}$$

The equality occurs if and only if each end is embedded. Here $m = 2$, because the total curvature of $\hat{M}'$ is -8π. Since $\chi(\hat{M}) \le 0$, also $\chi(\hat{M}') \le 0$ and we obtain

$$4 - 4k \ge -\chi(\hat{M}') \ge 0.$$

Hence, $k = 1$ and so $\chi(\hat{M}) = 0$. Since the equality holds, the immersed surface has only one end which is embedded. Therefore, by (2.18), the immersed surface is a plane, thus contradicting the hypothesis of nonorientability. This proves that does not exist such an immersed surface in R^3 with total curvature -4π.

Consider now a complete minimal immersion $x\colon \hat{M}-\{p_1,\dots,p_k\} \to R^3$ of a nonorientable surface, with total curvature -6π. Again, the formula of Jorge-Meeks applied to the oriented two-sheeted covering $\hat{M}'$ of $\hat{M}$ yields

$$6 \le 4k - \chi(\hat{M}')$$

because now $m = 3$. Hence,

$$3 - 2k \ge -\chi(\hat{M}).$$

Since the total curvature is -6π, the previous theorem implies that either $\chi(\hat{M}) = 1$ with $k \leq 2$ or $\chi(\hat{M}) = -1$ with $k = 1$. The last alternative is not acceptable, because then the equality holds, hence the surface has only one end which is embedded and so it is a plane, thus contradicting the nonorientability. It follows that $\hat{M}$ is a projective plane and the immersed surface is a projective plane minus k points, where either $k = 1$ or $k = 2$. To complete the proof we use the following

(2.25) PROPOSITION. There does not exist a complete minimal immersion $x: M \to R^3$ with total curvature -6π, where $M = \mathbb{P}^2 - \{p_1, p_2\}$.

Such a result excludes the case of $k = 2$. Therefore, we must have $\chi(\hat{M}) = 1$ and $k = 1$ and so the immersed surface is a projective plane minus one point; that is, M is a complete Möbius strip minimally immersed in R^3 with total curvature -6π. This concludes the proof.

The proof of above Proposition (2.25) is given in Chapter V.

CHAPTER IV

RECENT EXAMPLES OF COMPLETE MINIMAL SURFACES

Jorge-Meeks' examples of complete minimal surfaces in R^3, of genus zero with n ends, are presented in Section 1. These examples are immersed but have embedded ends.

In Section 2, we construct Meeks and Maria Elisa Oliveira's examples of complete minimal Möbius strips in R^3 with total curvature $-2m\pi$, $m \geq 3$. By the classification theorem of the previous chapter, the example corresponding to $m = 3$ is the unique complete minimal surface immersed in R^3 with total curvature -6π. Section 3 contains Oliveira's example of a nonorientable complete minimal surface of genus one with two ends and total curvature -10π.

In Section 4 we describe Klotz-Sario's examples of minimal immersions of surfaces with arbitrary genus and any number of ends.

Chen-Gackstätter's example of a complete minimal surface in R^3 of genus two with one end and total curvature -12π is presented in Section 5.

In Section 6, we exhibit another example of Chen-Gackstätter, namely, a complete minimal surface in R^3 of genus one with one end and total curvature -8π.

Section 7 is concerned with the construction of Costa's example of a complete minimal surface embedded in R^3 with total curvature -12π, of genus one with 3 ends. Two of such ends are of "catenoid type" and one is bounded. Costa has described this example in

1982 in his Doctoral dissertation at IMPA and has observed that it has embedded ends. It was only recently that D.A. Hoffman and W.H. Meeks proved that the whole surface is embedded. We present the construction of this example in two distinct ways. First, we do it by Chen-Gackstätter's method, starting from a suitable hyperelliptic Riemann surface. The second construction (Costa's method) is done by using the classical Weierstrass elliptic $\wp$-function on the plane.

Another Costa's example is exhibited in Section 8. It is a complete minimal surface of total curvature -20π, genus one and two ends, and its construction is accomplished by using the Weierstrass elliptic $\wp$-function.

The last section presents Jorge-Xavier's example of a complete minimal surface in R^3 lying between two parallel planes.

1. Complete minimal surfaces of genus zero with n ends.

For each integer $n \geq 1$, Jorge and Meeks [1] constructed an example of a complete minimal surface in R^3 with total curvature $-4\pi n$, conformally equivalent to $S^2 - \{p_1,\dots,p_{n+1}\}$, whose ends are embedded.

Take $\hat{M} = S^2(1)$ and identify $\hat{M}$ with $\mathbb{C} \cup \{\infty\}$ through the stereographic projection. Given $n \geq 1$, set $M = \hat{M} - \{z \in \mathbb{C};\ z^{n+1} = 1\}$ and define

$$(1.1)\qquad g(z) = \begin{cases} z^n, & \text{if } z \neq \infty, \\ \infty, & \text{if } z = \infty \end{cases} \qquad w = \begin{cases} \dfrac{dz}{(z^{n+1}-1)^2}, & \text{if } z \neq \infty \\ 0, & \text{if } z = \infty \end{cases}$$

For $z \neq \infty$, we obtain the 1-forms

(1.2)
$$\alpha_1 = \frac{1 - z^{2n}}{2(z^{n+1}-1)^2}\, dz ,$$
$$\alpha_2 = \frac{i(1+z^{2n})}{2(z^{n+1}-1)^2}\, dz ,$$
$$\alpha_3 = \frac{z^n}{(z^{n+1}-1)^2}\, dz .$$

At $z = \infty$ these forms vanish.

We want to apply Theorem (II 1.16) to obtain a minimal immersion

$$x = (x_1, x_2, x_3) \colon M \to R^3 ,$$

where

$$x_i = \int^z \alpha_i ,$$

and whose Gauss mapping is g. For this, we must show that α_k have no real periods on M. Since the poles of α_i are the $(n+1)^{th}$ roots of unity, this is equivalent to prove that the functions

(1.3)
$$F_1(z) = \frac{1-z^{2n}}{2(z^{n+1}-1)^2} ,$$
$$F_2(z) = \frac{i(1+z^{2n})}{2(z^{n+1}-1)^2} ,$$
$$F_3(z) = \frac{z^n}{(z^{n+1}-1)^2} ,$$

have real residues at $z = \theta$, when $\theta^{n+1} = 1$.

I) Computation of the residue of F_1 at $z = \theta$.

1st case: $\theta = \pm 1$. Observe that $\theta = \pm 1$ occurs if and only if $\theta^{2n} = 1$. Then

$$1-z^{2n} = \theta^{2n}-z^{2n} = (\theta-z)\sum_{j=1}^{2n} \theta^{2n-j} z^{j-1},$$

$$z^{n+1}-1 = z^{n+1}-\theta^{n+1} = (z-\theta)\sum_{j=0}^{n} \theta^{n-j} z^{j}$$

and

$$F_1(z) = -\frac{1}{2}\left(\frac{1}{(z-\theta)} \frac{\sum_{j=1}^{2n} \theta^{2n-j} z^{j-1}}{\left(\sum_{j=0}^{n} \theta^{n-j} z^{j}\right)^2}\right) = -\frac{1}{2}\frac{1}{z-\theta} G_1(z) .$$

Now, if $z = \theta$, we obtain

$$G_1(\theta) = \frac{\sum_{j=1}^{2n} \theta^{2n-j}\theta^{j-1}}{\left(\sum_{j=0}^{n} \theta^{n-j}\theta^{j}\right)^2} = \frac{\sum_{j=1}^{2n} \theta^{2n-1}}{\left(\sum_{j=0}^{n} \theta^{n}\right)^2} = \frac{2n\,\theta^{2n-1}}{((n+1)\theta^{n})^2} = \frac{2n}{(n+1)^2\theta} .$$

Hence, $z = \theta$ is a pole of order one of F_1. Thus,

(1.4) $$\operatorname*{Res}_{z=\pm 1} F_1 = -\frac{1}{2}\lim_{z\to\pm 1}\left(\frac{2n}{(n+1)^2 z}\right) = \frac{\pm n}{(n+1)^2} .$$

2nd case: $\theta \neq \pm 1$. We may then rewrite F_1 as

$$F_1 = \frac{1}{2(z-\theta)^2}\frac{1-z^{2n}}{\left(\sum_{j=0}^{n} \theta^{n-j} z^{j}\right)^2} .$$

Observe that $\frac{1-\theta^{2n}}{(n+1)^2\theta^{2n}}$ is finite and nonzero. Hence, $z = \theta$ is a pole of order 2 of F_1. It follows that

$$\operatorname*{Res}_{z=\theta} F_1 = \frac{d}{dz}\left[\frac{1}{2}\,\frac{1-z^{2n}}{\left(\sum\limits_{j=0}^{n} \theta^{n-j}z^j\right)^2}\right]_{z=\theta} =$$

$$= \left[\frac{-2n\ z^{2n-1}}{2\left(\sum\limits_{j=0}^{n} \theta^{n-j}z^j\right)^2} - \frac{2(1-z^{2n})\left(\sum\limits_{j=1}^{n} j\ \theta^{n-j}z^{j-1}\right)}{2\left(\sum\limits_{j=0}^{n} \theta^{n-j}z^j\right)^3}\right]_{z=\theta} .$$

Thus,

$$\operatorname*{Res}_{z=\theta} F_1 = -\frac{2n\ \theta^{2n-1}}{2(n+1)^2\theta^{2n}} - \frac{2(1-\theta^{2n})\left(\frac{n(n+1)}{2}\ \theta^{n-1}\right)}{2(n+1)^3\ \theta^{3n}} = \tag{1.5}$$

$$= -\frac{n\theta^{-1}}{(n+1)^2} - \frac{n(1-\theta^{2n})\theta^{n-1}}{2(n+1)^2\theta^{3n}} =$$

$$= -\frac{n}{2(n+1)^2}\left(2\theta^{-1} + \frac{1-\theta^{2n}}{\theta^{2n+1}}\right) = -\frac{n}{2(n+1)^2}\,\frac{\theta^{2n}+1}{\theta^{2n+1}} .$$

Since $\theta^{n+1} = 1$, $\theta^n = \theta^{-1}$. Thus,

$$\operatorname*{Res}_{z=\theta} F_1 = -\frac{n}{2(n+1)^2}(\theta+\bar{\theta}) = \frac{-n\ \mathrm{Re}\ \theta}{(n+1)^2}.$$

Therefore, the residues of F_1 are all real.

II) <u>Computation of the residue of</u> F_2 <u>at</u> $z = \theta$.

The computation is done following the same lines. We will consider also two cases: $\theta = \pm i$, which are poles of order one with $\operatorname*{Res}_{z=\pm i} F_2 = \pm n/(n+1)^2$; and $\theta \neq \pm i$, which are poles of order two with $\operatorname*{Res}_{z=\theta} F_2 = n(\mathrm{Im}\ \theta)/(n+1)^2$. Therefore, in both cases, the residues of F_2 are real.

III) Computation of the residue of F_3 at $z = \theta$.

Observe that

$$F_3 = \frac{z^n}{(z^{n+1} - 1)^2} = -\frac{1}{n+1}\frac{d}{dz}\left(\frac{1}{z^{n+1}-1}\right).$$

Thus,

(1.6) $\operatorname{Res}_{z=\theta} F_3 = 0$, for each θ that satisfies $\theta^{n+1} = 1$.

From I, II and III we conclude that the α_i have no real periods on M. Now we will show that M with the induced metric is complete. The metric on M is given by

$$ds = \frac{1+|z^n|^2}{|z^{n+1}-1|^2\sqrt{2}}\,|dz|.$$

We have to show that if γ is a divergent path on M (that is, if γ is a path on $\mathbb{C}$ that has one of the roots of unity as an end point), then

$$\sqrt{2}\,L(\gamma) = \int_\gamma \frac{1+|z^n|^2}{|z^{n+1}-1|^2}\,|dz| = \infty.$$

In fact, consider such a $\gamma: [0,L) \to M$ parametrized by arc length and write $\gamma(s)^{n+1} = 1 + re^{i\alpha}$. Then $\lim_{s\to L} r = 0$ and

$$(n+1)|\gamma|^n = |r' + ir\alpha'| \geq |r'|.$$

For s sufficiently close to L, $|\gamma|$ is close to 1 and $(n+1)|\gamma|^n \leq A$. Then,

$$(1.7)\qquad L(\gamma) \geq c_1 + c_2 \int_a^L \frac{ds}{r^2} \geq c_1 + \frac{c_2}{A}\int_a^L \frac{|r'|\,ds}{r^2} =$$

$$= c_1 + \frac{c_2}{A}\int_b^0 \frac{-dr}{r^2} = \infty .$$

Finally, the total curvature $C(M)$ of M is $-4\pi n$. This is a simple consequence of the fact that $g(z) = z^n$ covers $S^2(1)$ n times. On the other hand, we know from Theorem (III 2.11) that, if M has $n+1$ ends, then

$$C(M) = 2\pi(\chi(M)-(n+1))$$

occurs if and only if each end of M is embedded. In our example $\chi(M) = 2-(n+1)$. Hence the above equality is satisfied and the $n+1$ ends of M are embedded.

Summarizing, we have proved the following

(1.8) THEOREM. <u>For each integer</u> $n \geq 1$ <u>there exists a complete surface minimally immersed in</u> R^3 <u>with total curvature</u> $-4\pi n$, <u>conformally equivalent to</u> $S^2 - \{p_1,\dots,p_{n+1}\}$, <u>whose ends are embedded. Its Weierstrass representation is given by</u> (1.1).

OBSERVATIONS. 1) For each $n \geq 1$, the example just constructed is invariant by a rotation through the angle $2\pi/(n+1)$ around the z-axis.

2) We know that for $n = 1$ the above example is the catenoid, which is embedded. For $n > 1$ the examples obtained have embedded ends, but they are not embedded.

The pictures in the next page show one of the above examples with three ends.

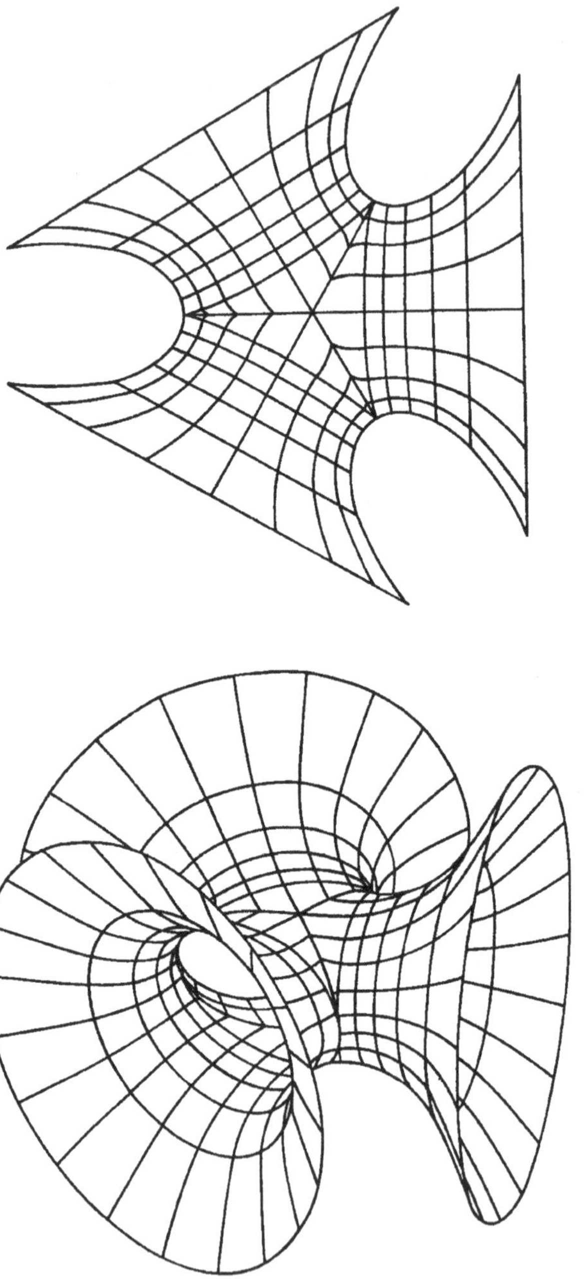

Fig. 8

Two views of a Jorge-Meeks surface

2. Complete minimal Möbius strips.

In this section we present examples of complete Möbius strips in R^3, with total curvature $-2\pi m$, for any odd integer $m \geq 3$. These examples have been obtained by Meeks [1] (case $m=3$) and Oliveira [1] (cases $m > 3$). Take $\tilde{M} = \mathbb{C}-\{0\}$, and define $g:\tilde{M} \to \mathbb{C} \cup \{\infty\}$ and ω by

$$(2.1) \qquad g(z) = \begin{cases} z^{m-1} \dfrac{z+1}{z-1}, & z \neq 1 \\ \infty, & z = 1 \end{cases}$$

$$\omega = i \frac{(z-1)^2}{z^{m+1}} dz ,$$

where m is an odd integer greater than or equal to three. By using (II 1.19) we obtain

$$\alpha_1 = \frac{i}{2} \left[\frac{(z-1)^2}{z^{m+1}} - z^{m-3}(z+1)^2\right] dz ,$$

$$(2.2) \qquad \alpha_2 = -\frac{1}{2} \left[\frac{(z-1)^2}{z^{m+1}} + z^{m-3}(z+1)^2\right] dz ,$$

$$\alpha_3 = i \frac{z^2-1}{z^2} dz .$$

It is straightforward to verify that α_1, α_2 and α_3 do not have real periods thus giving rise to a minimal immersion $\tilde{x}: \tilde{M} \to R^3$. The induced metric on $\tilde{M}$ is

$$(2.3) \qquad ds = \frac{1}{\sqrt{2}} \left(\frac{|z-1|^2}{|z|^{m+1}} + |z|^{m-3}|z+1|^2\right)|dz| .$$

If γ is a divergent path in $\tilde{M}$, it is easy to verify that

$$L(\gamma) = \int_{\gamma} ds = \infty \ ;$$

Thus, in the induced metric, $\tilde{M}$ is complete. Furthermore, g extends to a meromorphic mapping

$$\tilde{g}: \mathbb{C} \cup \{\infty\} \to \mathbb{C} \cup \{\infty\}$$

that covers each point of $\mathbb{C} \cup \{\infty\}$ m times. As a consequence, the total curvature of $\tilde{M}$ is $-4\pi m$.

When we identify $S^2(1)$ with $\mathbb{C} \cup \{\infty\}$ the antipodal mapping corresponds to the transformation $I(z) = -1/\bar{z}$. The projective space $\mathbb{P}^2$ may then be identified with the quotient of $\mathbb{C} \cup \{\infty\}$ by the equivalence relation "$z \sim w$ if and only if $w = I(z)$". Let $\pi: \mathbb{C} \cup \{\infty\} \to \mathbb{P}^2$ be the canonical projection that takes each point into its equivalence class. It is easy to see that $\mathbb{P}^2 - \pi(z_0)$ is a Möbius strip for any choice of z_0. Set $M = \mathbb{P}^2 - \pi(0)$. We are going to show that the immersion $\tilde{x}: \tilde{M} \to R^3$ can be factored out by I; that is, there exists an immersion $x: M \to R^3$ such that $\tilde{x} = x \circ \pi$. For this to happen it is necessary and sufficient that

$$\tilde{x}(I(z)) = \tilde{x}(z), \tag{2.4}$$

for each z in $\mathbb{C}-\{0\} = \pi^{-1}(M)$. If $\alpha_j(z) = \phi_j(z)dz$, it follows from (2.2) that $\phi_j(I(z)) = \bar{z}^2 \overline{\phi_j(z)}$. As a consequence, we have $I^*\alpha_j = \bar{\alpha}_j$, hence

$$(2.5)\qquad \tilde{x}_j(I(z)) = \mathrm{Re}\int_{z_o}^{I(z)} \alpha_j = \mathrm{Re}\Big(\int_{z_o}^{I(z_o)} \alpha_j + \int_{I(z_o)}^{I(z)} \alpha_j\Big) =$$

$$= w_j + \mathrm{Re}\int_{z_o}^{z} I^*\alpha_j = w_j + \mathrm{Re}\int_{z_o}^{z} \bar{\alpha}_j =$$

$$= w_j + \mathrm{Re}\int_{z_o}^{z} \alpha_j = w_j + \tilde{x}_j(z)\ .$$

Since $\tilde{x}_j(z) = \tilde{x}_j(I(I(z))) = w_j + \tilde{x}_j(I(z)) = 2w_j + \tilde{x}_j(z)$, $w_j = 0$. Hence, $\tilde{x}_j(I(z)) = \tilde{x}_j(z)$, for all z and $j = 1,2,3$. Thus, (2.4) is true. Therefore, the mapping $x\colon M \to R^3$ is well defined.

Since $\pi\colon \tilde{M} \to M$ is the oriented two-sheeted covering for M, we may define a metric in M in such way that π becomes a local isometry. The mapping x is then an isometric minimal immersion. Since $\tilde{M}$ is complete, so is M, and this completes the example. Observe that x defines a complete minimal immersion of the Möbius strip into R^3; but, since $M = \mathbb{P}^2 - \pi(0)$, the immersion has one end. The total curvature of M must be half of that of $\tilde{M}$. Therefore, the total curvature of $\tilde{M}$ is $-2\pi m$.

In Chapter V we will prove that for $m = 3$ this example is the unique complete Möbius strip minimally immersed in R^3 with total curvature -6π (cfr. Proposition 1.34 of that chapter).

3. A nonorientable complete minimal surface of genus one with two ends and total curvature -10π

In this section we present the example obtained by Maria Elisa G.G. Oliveira [1] of a nonorientable complete minimal surface of genus one with two ends and total curvature -10π.

Take $\tilde{M} = \mathbb{C}-\{0,1,-1\}$, $g\colon \tilde{M} \to \mathbb{C} \cup \{\infty\}$ and ω defined by

(3.1)

$$g(z) = \begin{cases} \dfrac{z^3(z^2-b^2)}{b^2z^2-1}, & z^2 \neq 1/b^2, \\[2ex] \infty, & z^2 = 1/b^2, \end{cases}$$

$$\omega = \frac{i(b^2z^2-1)^2}{z^2(z-1)^4(z+1)^4}\, dz,$$

where b is a real constant. Making use of (II 1.19), we obtain

(3.2)

$$\alpha_1 = \frac{i}{2}\,\frac{(b^2z^2-1)^2-z^6(z^2-b^2)^2}{z^2(z-1)^4(z+1)^4}\, dz = \frac{P_1(z)}{z^2(z-1)^4(z+1)^4}\, dz,$$

$$\alpha_2 = -\frac{1}{2}\,\frac{(b^2z^2-1)^2+z^6(z^2-b^2)^2}{z^2(z-1)^4(z+1)^4}\, dz = \frac{P_2(z)}{z^2(z-1)^4(z+1)^4}\, dz,$$

$$\alpha_3 = \frac{iz^3(z^2-b^2(b^2z^2-1)}{z^2(z-1)^4(z+1)^4}\, dz = \frac{P_3(z)}{z^2(z-1)^4(z+1)^4}\, dz.$$

The constant b is to be determined in such a way that the forms α_j have no real periods. The computation of the periods can be done by using Cauchy's integral formula. We obtain

$$(3.3)\qquad \int_{\gamma_0} \frac{P_j(z)dz}{z^2(z-1)^4(z+1)^4} = 2\pi i\ P'_j(0) = 0,$$

for any small closed curve γ_0 around $z = 0$. To examine what happens around $z = 1$, we take derivatives of $f_j(z) = P_j(z)/z^2(z+1)^4$ to obtain

$$f'''_j(1) = \frac{1}{16}(P'''_j(1) - 12\ P''_j(1) + 57\ P'_j(1) - 105\ P(1)).$$

Of course, for a small curve γ_1 around $z = 1$, we have

$$(3.4)\qquad \int_{\gamma_1} \frac{P_j(z)dz}{z^2(z-1)^4(z+1)^4} = \frac{2\pi i}{3!}\ f'''_j(1).$$

It is a straightforward computation to show that $f'''_2(1) = f'''_3(1) = 0$, and that

$$(3.5)\qquad f'''_1(1) = \frac{i}{32}(6b^4+60b^2-210)\ .$$

Now, we determine b by solving the equation $f'''_1(1) = 0$. One finds $b = -5 \pm 2\sqrt{15}$. With this choice of b the forms α_1, α_2 and α_3 have no periods and give rise to a minimal immersion $\tilde{x}\colon \tilde{M} \to R^3$. It is not necessary to verify the existence of real periods around $z = -1$. We just have to observe that, if $I(z) = -1/\bar{z}$ and $\alpha_j = \phi_j(z)dz$, then, for the above choice of b, we have

$$(3.6)\qquad \phi_j(I(z)) = \bar{z}^2\ \overline{\phi_j(z)}\ .$$

Since I is a conformal diffeomorphism, a small curve γ_1 around $z = 1$ will be transformed into a small curve $\gamma_2 = I(\gamma_1)$ around $z = -1$ and

$$\int_{\gamma_2} \alpha_j = \int_{\gamma_1} I^*\alpha_j = \int_{\gamma_1} \bar{\alpha}_j = \overline{\int_{\gamma_1} \alpha_j} = 0 .$$

The induced metric on $\tilde{M}$ is

$$(3.7) \qquad ds = \frac{1}{\sqrt{2}} \left(\frac{|b^2z^2-1|^2}{|z|^2|z^2-1|^4} + \frac{|z|^4|z^2-b^2|^2}{|z^2-1|^4} \right) |dz| .$$

If γ is a divergent path in $\tilde{M}$, it is easy to verify that

$$L(\gamma) = \int_{\gamma} ds = \infty$$

and then, in the induced metric, $\tilde{M}$ is complete. Furthermore, g extends to a meromorphic function

$$\tilde{g}: \mathbb{C} \cup \{\infty\} \to \mathbb{C} \cup \{\infty\}$$

which covers each point of $\mathbb{C} \cup \{\infty\}$ five times. As a consequence of this, the total curvature of $\tilde{M}$ is -20π.

Taking into account (3.6) and doing the same argument as in (2.5) of the previous section, we conclude that

$$\tilde{x}(I(z)) = \tilde{x}(z).$$

Therefore, $\tilde{x}$ can be factored out by I; that is, there exists an immersion $x: M \to R^3$ such that $\tilde{x} = x \circ \pi$, where π is the canonical projection from the sphere into the real projective space $\mathbb{P}^2$ and $M = \mathbb{P}^2 - \pi\{0,1\}$. The details are the same as those in the previous section. Since $\pi: \tilde{M} \to M$ is the two-sheeted covering for M, we may define the metric in M in such a way that π becomes a local isometry. The mapping x is then an isometric minimal immersion. Since $\tilde{M}$ is complete, then M is also complete. Observe that $\tilde{x}$ defines a minimal immersion with four ends. Hence, x have just two ends. The total curvature of M must be half of the total curvature of $\tilde{M}$; therefore, it is -10π. Observe that M can be considered as a Möbius strip minus one point.

4. Complete minimal immersions of surfaces of arbitrary genus with any number of ends.

In this section we are going to prove a theorem due to Klotz and Sario [1].

(4.1) THEOREM. There exists a complete minimal immersion $x: M \to R^3$ of a surface of genus g, $M = \hat{M} - \{p_1,\ldots,p_k\}$, where $\hat{M}$ is a compact surface such that:

a) $\chi(\hat{M}) = 2$ and $k \geq 1$, or

b) $\chi(\hat{M}) = (2-2g) < 2$ and $k \geq 4$.

The proof of this theorem is based on the simple idea that, if $Y: M_0 \to R^3$ is a complete minimal immersion and $\pi: M \to M_0$ is a covering for M_0, then $Y\circ\pi: M \to R^3$ is also a complete minimal immersion. Klotz and Sario have started with a complete minimal immersion Y from $M_0 = S^2(1) - \{p_1,p_2,p_3\}$ into R^3 and have shown that it is possible to cover M_0 by a surface $\hat{M} - \{p_1,\ldots,p_k\}$, under either one of the conditions (a) or (b) above. The immersions $x = Y\circ\pi$ so obtained are complete, minimal and their domains may have a very complicated topology, but they all have the same image. This justifies the effort in trying to obtain geometrically distinct examples of complete minimal surfaces in R^3 with arbitrary topological type, as, for example, was done by Jorge, Meeks and Oliviera. In any case, this theorem leaves open questions about the existence of complete minimal immersions $x: M \to R^3$ where $M = \hat{M} - \{p_1,\ldots,p_k\}$ with

$\chi(\hat{M}) < 2$ and $k = 1,2$ and 3.

Once and for all, in this section, set $M_o = S^2(1) - \{p_1,p_2,p_3\}$ and let $Y\colon M_o \to R^3$ be the complete minimal immersion described in Section 1 of this chapter, where p_1, p_2 and p_3 are the cubic roots of unity. The proof of the theorem depends on the following two lemmas.

(4.2) LEMMA. <u>For any integer</u> $k \geq 3$ <u>there exists a covering projection</u> $\pi\colon S^2(1) - \{q_1,\dots,a_k\} \to M_o$.

<u>Proof</u>: To prove this lemma, take $k-2$ copies of M_o and cut all of them along a path α connecting p_1 to p_2. Open each copy of M_o along the slit α. The closure of each one of them contains, now, two copies of the path α. Call one of them α_1 and the other one α_2. Identify the points of α_1 of the j-copy with the points of α_2 of the $(j+1)$-copy, $1 \leq j \leq k-3$; then, identify the points of α_1 of the $(k-2)$-copy with the points of α_2 of the first copy. The surface M_k so constructed is conformally equivalent to $S^2(1) - \{q_1,\dots,q_k\}$. For the reader to see easily this fact, imagine $S^2(1)$ open along the slit, as a rectangle with sides identified as indicated in the following picture:

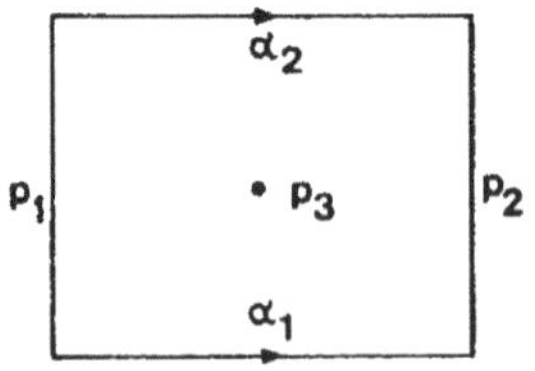

Fig. 9

It is then clear that M_k, obtained from the identification of $k-2$ copies of this rectangle, as described above, will be topologically a rectangle with sides identified as indicated in the picture below.

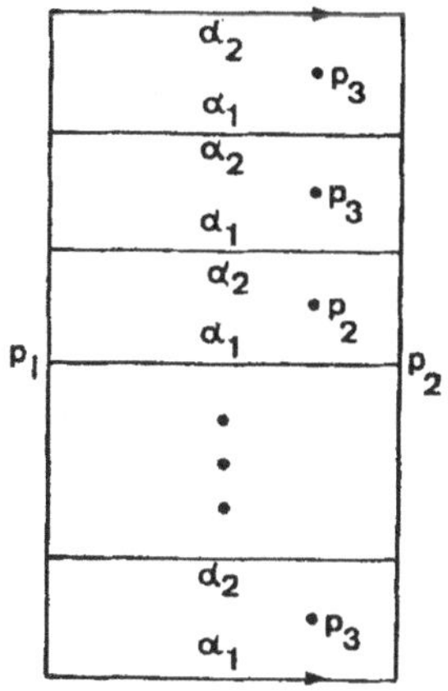

Fig. 10

Observe that, besides p_1 and p_2, there is one missing point on each copy used in the construction of M_k. Hence, it is obvious that M_k is a sphere minus k points. The desired covering projection $\pi: M_k \to M_0$ is the mapping which carries each point of M_k to the point of M_0 that has given origin to it.

(4.3) LEMMA. <u>For any integers</u> $g > 0$ <u>and</u> $n \geq 2(g+1)$, <u>there is a covering</u> $\pi: \hat{M} - \{q_1,\ldots,q_k\} \to M_0$ <u>with</u> $\chi(\hat{M}) = 2-2g$ <u>and either</u> $k = 2(n-g-1)$ <u>or</u> $k = 2(n-g-1) + 1$.

<u>Proof</u>: Given integers $g > 0$ and $n \geq 2(g+1)$, consider two copies of $M_n = S^2(1) - \{p_1,\ldots,p_n\}$. Cut $S^2(1)$ along simple curves α_i connecting p_{2i-1} to p_{2i}, $1 \leq i \leq g+1$, in such way that α_i and α_j, $i \neq j$, have no common points. Each copy must be open along the slits α_i. The closure of each one contains two copies of each path

α_i. Call one of them α_{i1} and the other one α_{i2}. Each copy of M_n, after the slit, will be, topologically, the surface of the picture below; that is, a sphere minus $(g+1)$ disks and $n-(2g+2)$ points.

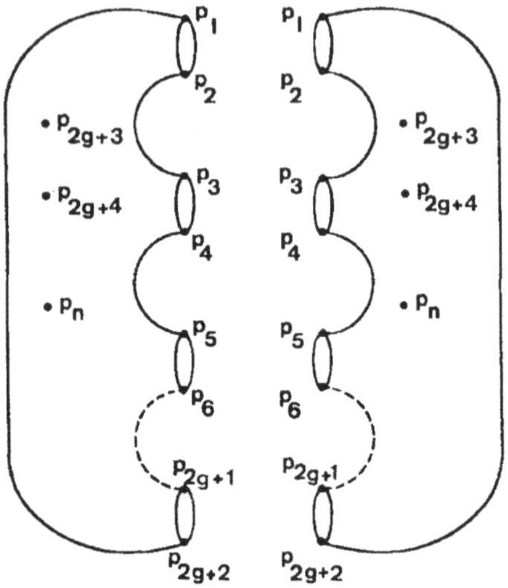

Fig. 11

Now, we identify the points of α_{i1} of the first copy with the points of α_{i2} of the second copy and the points of α_{i2} of the first copy with the points of α_{i1} of the second copy. This is equivalent to identify the boundaries of the corresponding missing disks of the two copies. The resulting surface, M_{gk}, is, topologically, a sphere with g handles from which $k = 2(n-2g-2) + 2g + 2 = 2(n-g-1)$ points have been removed. The covering projection $\pi: M_{gk} \to M_o$ is the mapping which carries each point of M_{gk} to the point of M_o that has given origen to it.

If we had started with $M_n = S^2(1) - \{p_1,\ldots,p_n\}$ and

$M_{n+1} = S^2(1) - \{p_1,\ldots,p_n,p_{n+1}\}$, and followed the same reasoning, we would had ended with a covering $\pi: M_{gk} \to M_o$, where M_{gk} would have g handles and $k = 2(n-g-1) + 1$ points removed. This proves the lemma.

<u>Proof of the theorem</u>: We first observe that the simplest covering obtained by this procedure is $\pi: M_{14} \to M_o$. Let us take $g+1$ copies of M_{14}. As before, cut each one of them along two curves α and β connecting respectively p_1 to p_2 and p_3 to p_4, and open the surfaces along the slits. The closure of each one of them contains, now, two copies of the path α and two copies of the path β. Call them α_1, α_2, β_1 and β_2. Each one of the slit copy of M_{14} is now, topologically, two rectangles connected by a tube, as in the picture below.

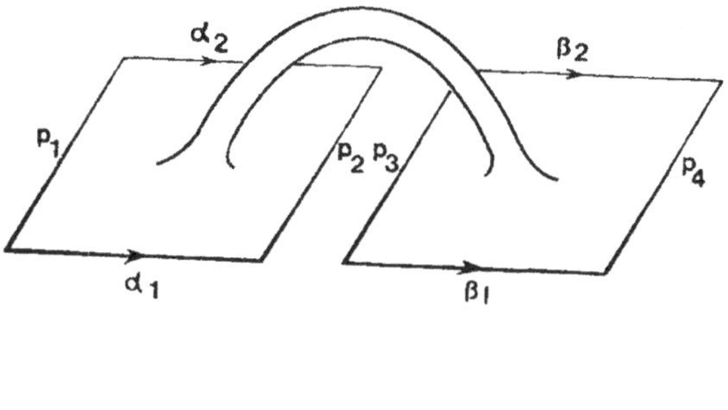

Fig. 12

We now past these $g+1$ identical surfaces along the slits by identification of the points of α_1 and β_1 of the j-copy with, respectively, the points of α_2 and β_2 of the $(j+1)$-copy, $1 \le j \le g$,

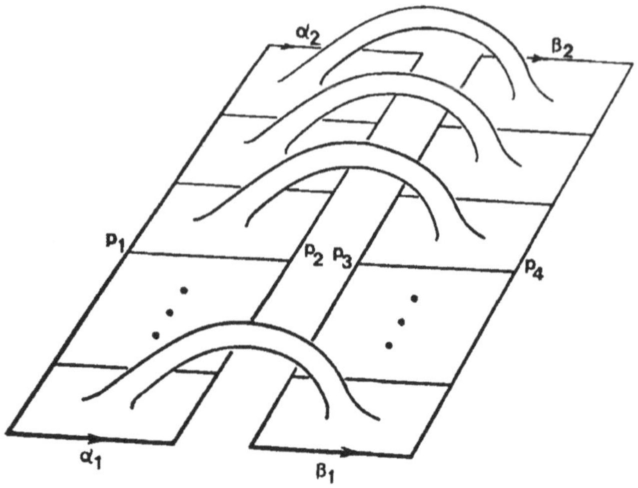

Fig. 13

and the points of α_1 and β_1 of the last one with the α_2 and β_2 of the first one. The resulting surface, M_{g4}, consists of two spheres connected by $g+1$ tubes from which four points were removed, namely, a sphere with g handles minus four points.

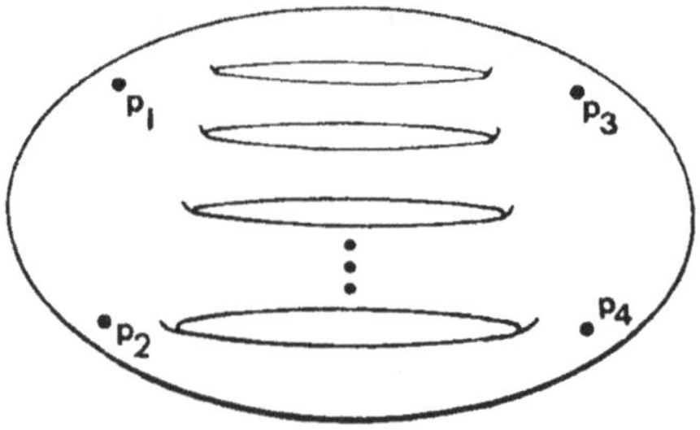

Fig. 14

If, in the above construction, we had pasted g slit copies of M_{14} and one copy of M_{1k}, $k \geq 4$, we would had obtained a surface M_{gk}, $k \geq 4$, which would be topologically a sphere with g handles with k points removed. This proves the theorem.

5. A complete minimal surface of genus two with one end and total curvature -12π.

The example that will be described in this section was discovered by Chen and Gackstätter [1]. Let $\hat{M}$ be a Riemann surface where the function $w(z)$, given by

$$w^2(z) = z(z^2-a^2)(z^2-b^2), \tag{5.1}$$

is well defined. We will assume that a and b are real constants, with $b > a > 0$. The surface $\hat{M}$ is obtained by cutting the sphere $\mathbb{C} \cup \{\infty\}$ along 3 curves connecting $-b$ to $-a$, 0 to a and b to ∞, and then by pasting two copies of this slit sphere along the slits, as was done in the last section. The resulting surface $\hat{M}$ is then a sphere with two handles.

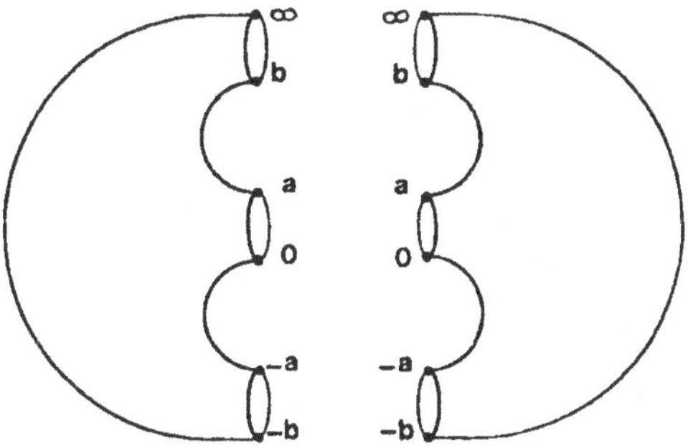

Fig. 15

Take $M = \hat{M} - \{\infty\}$, $g: M \to \mathbb{C} \cup \{\infty\}$ defined by

(5.2)
$$g(z) = B\, w(z)/(z^2-a^2) \quad \text{and}$$
$$\omega = (z^2-a^2)dz/w(z),$$

where B is a constant. A way of studying a function $F(z,w(z))$ (or a 1-form $\Gamma = F(z,w(z))dz$) around a branch point z_0 of the function w is to consider a new parameter ξ, given by $\xi^2 = z-z_0$, and to study the function $F(\xi^2+z_0, w(\xi^2+z_0))$ (or the 1-form $\tilde{\Gamma} = F(\xi^2+z_0, w(\xi^2+z_0))2\xi d\xi$). It is then easy to verify that, at the points $z = a$ and $z = -a$, g has poles of order one and ω has zeroes of order two, and that $z = 0$, $z = b$ and $z = -b$ are regular points of the 1-form ω. It follows from (5.2) that

(5.3)
$$\alpha_1 = \frac{1}{2}\left(\frac{z^2-a^2}{w} - B^2 \frac{w}{z^2-a^2}\right)dz ,$$
$$\alpha_2 = \frac{i}{2}\left(\frac{z^2-a^2}{w} + B^2 \frac{w}{z^2-a^2}\right)dz ,$$
$$\alpha_3 = B\, dz .$$

Such forms have no real periods. The existence of real periods should be searched among the cycles that generate the fundamental group of M. The surface M has fundamental group generated by four homotopy classes. We may choose one curve in each one of these classes as indicated in the picture below, where the dark lines and the light lines stand for the trajectories on different copies of the slit plane used

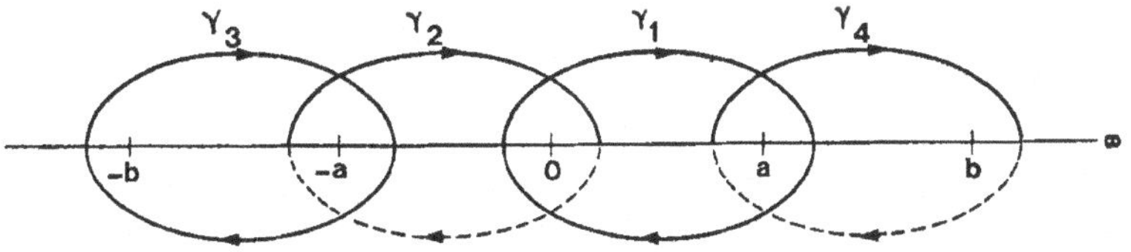

Fig. 16

to construct $\hat{M}$. The next picture indicates how the curves γ_1, γ_2, γ_3 and γ_4 lie in $\hat{M}$.

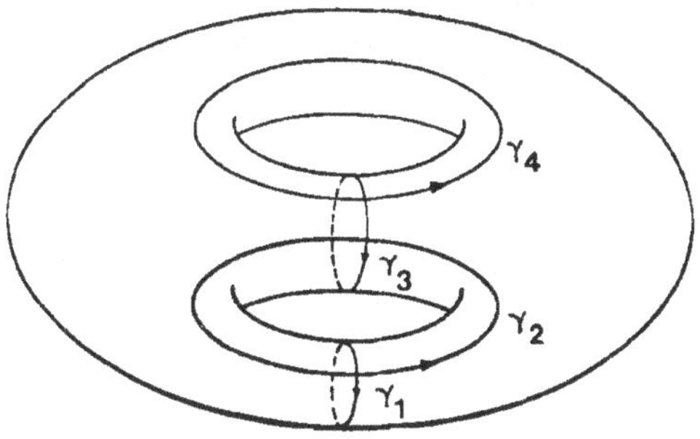

Fig. 17

Since α_3 is an exact form we have

(5.5)
$$\int_{\gamma_k} \alpha_3 = 0, \qquad 1 \le k \le 4.$$

To compute $\int_{\gamma_k} \alpha_j$, $1 \le j \le 2$, $1 \le k \le 4$, it is convenient to compute first the values of $\int_{\gamma_k} f\,dz$ and $\int_{\gamma_k} (1/f)dz$, for $f = w/(z^2-a^2)$. We obtain

$$\int_{\gamma_1} f dz = \int_0^a \frac{x(x^2-b^2)dx}{\sqrt{x(x^2-a^2)(x^2-b^2)}} - \int_0^a \frac{x(x^2-b^2)dx}{-\sqrt{x(x^2-a^2)(x^2-b^2)}} \quad \text{and}$$

(5.6)

$$\int_{\gamma_1} (1/f)dz = \int_0^a \frac{(x^2-a^2)dx}{\sqrt{x(x^2-a^2)(x^2-b^2)}} - \int_0^a \frac{(x^2-a^2)dx}{-\sqrt{x(x^2-a^2)(x^2-b^2)}} ,$$

where the minus signs, in the second integrals, comes from the orientation of the curve and from the chosen branch of the square root. Thus,

$$\int_{\gamma_1} f\,dz = -2 \int_0^a \frac{x(b^2-x^2)dx}{\sqrt{x(x^2-a^2)(x^2-b^2)}} = -2\,F_1 ,$$

(5.7)

$$\int_{\gamma_1} (1/f)dz = -2 \int_0^a \frac{(a^2-x^2)dx}{\sqrt{x(x^2-a^2)(x^2-b^2)}} = -2\,G_1 .$$

In a similar way we obtain

$$\int_{\gamma_2} f\,dz = -2iF_1 , \qquad \int_{\gamma_2} (1/f)dz = 2iG_1 ,$$

(5.8)

$$\int_{\gamma_3} f\,dz = 2\,F_2 , \qquad \int_{\gamma_3} (1/f)dz = 2\,G_2 ,$$

$$\int_{\gamma_4} f\,dz = 2i\,F_2 , \qquad \int_{\gamma_4} (1/f)dz = -2iG_2 ,$$

where

$$F_2 = \int_a^b \frac{x(b^2-x^2)\,dx}{\sqrt{x(x^2-a^2)(b^2-x^2)}} \quad \text{and} \quad G_2 = \int_a^b \frac{(x^2-a^2)\,dx}{\sqrt{x(x^2-a^2)(b^2-x^2)}} .$$

Observe that F_1, F_2, G_1 and G_2 are positive constants which depend only on a and b. The table below gives the values of $\int_{\gamma_k} \alpha_j$.

(5.9)

	$\gamma=\gamma_1$	$\gamma=\gamma_2$	$\gamma=\gamma_3$	$\gamma=\gamma_4$
$\int_\gamma \alpha_1$	$-G_1+B^2F_1$	$i(G_1+B^2F_1)$	$G_2-B^2F_2$	$-i(G_2+B^2F_2)$
$\int_\gamma \alpha_2$	$-i(G_1+B^2F_1)$	$-G_1+B^2F_1$	$i(G_2+B^2F_2)$	$G_2-B^2F_2$

Thus, in order that real periods do not exist, we must have

$$G_1 = B^2F_1 \quad \text{and} \quad G_2 = B^2F_2 \tag{5.10}$$

or, equivalently,

$$G_1 = B^2F_1 , \tag{5.11}$$

$$F_1G_2 = G_1F_2 . \tag{5.12}$$

For fixed a and b, we may choose B in such way that (5.11) occurs. Let us then fix $a = 1$ and try to find $b > 1$ such that (5.12) is verified. That is, we must find $b > 1$ such that

$$\left(\int_0^1 \frac{(1-x^2)\,dx}{\sqrt{x(x^2-1)(x^2-b^2)}} \right) \left(\int_1^b \frac{x(b^2-x^2)\,dx}{\sqrt{x(x^2-1)(b^2-x^2)}} \right) =$$

$$= \left(\int_0^1 \frac{x(b^2-x^2)\,dx}{\sqrt{x(x^2-1)(x^2-b^2)}} \right) \left(\int_1^b \frac{(x^2-1)\,dx}{\sqrt{x(x^2-1)(b^2-x^2)}} \right).$$

From now on, we write for short LHS (RHS) for left hand side (right hand side).

Changing the parameter x for $x+1$ in the integrals from 1 to b, we obtain

$$\left(\int_0^1 \sqrt{\frac{1+x}{(b+x)(b-x)}} \sqrt{\frac{1-x}{x}}\, dx \right) \left(\int_0^{b-1} \sqrt{\frac{(1+x)(b+1+x)}{2+x}} \sqrt{\frac{b-1-x}{x}}\, dx \right) =$$

(5.13)

$$\left(\int_0^1 \sqrt{\frac{(b-1-x)(b+1-x)}{2-x}} \sqrt{\frac{1-x}{x}}\, dx \right) \left(\int_0^{b-1} \sqrt{\frac{b+1-x}{(b-x)(2b-x)}} \sqrt{\frac{b-1-x}{x}}\, dx \right).$$

Now, if $b \to \infty$, we have

$$\text{LHS of (5.13)} \le \left(\sqrt{\frac{2}{b(b-1)}} \int_0^1 \sqrt{\frac{1-x}{x}}\, dx \right) \left(\sqrt{2b} \int_0^{b-1} \sqrt{\frac{b-1-x}{x}}\, dx \right)$$

(5.14)

$$\text{RHS of (5.13)} \ge \left(\frac{b(b-1)}{2} \int_0^1 \frac{1-x}{x}\, dx \right) \left(\frac{1}{2b} \int_0^{b-1} \frac{b-1-x}{x}\, dx \right).$$

On the other hand, if b is close to 1, say, $b = 1+\epsilon$, $\epsilon > 0$, then

$$\text{LHS of (5.13)} = \left(\int_0^1 \sqrt{\frac{1+x}{(1+\epsilon+x)(1+\epsilon-x)}}\sqrt{\frac{1-x}{x}}\,dx\right)\left(\int_0^\epsilon \sqrt{\frac{(1+x)(2+\epsilon+x)}{2+x}}\sqrt{\frac{\epsilon-x}{x}}\,dx\right)$$

(5.15)

$$\geq \left(\int_0^1 \sqrt{\frac{1}{2+\epsilon}\cdot\frac{1}{1+\epsilon-x}}\sqrt{\frac{1-x}{x}}\,dx\right)\left(\int_0^\epsilon \sqrt{\frac{\epsilon-x}{x}}\,dx\right.$$

$$\text{RHS of (5.13)} = \left(\int_0^1 \sqrt{\frac{(\epsilon+x)(2+\epsilon-x)}{2-x}}\sqrt{\frac{1-x}{x}}\,dx\right)\left(\int_0^\epsilon \sqrt{\frac{2+\epsilon-x}{(1+\epsilon-x)(2+2\epsilon-x)}}\sqrt{\frac{\epsilon-x}{x}}\,dx\right)$$

(5.16)

$$\leq \left(\int_0^1 \sqrt{\frac{2+\epsilon}{2}}\sqrt{\epsilon+x}\sqrt{\frac{1-x}{x}}\,dx\right)\left(\int_0^\epsilon \sqrt{\frac{\epsilon-x}{x}}\,dx\right).$$

But,

$$\sqrt{\frac{1}{2+\epsilon}\cdot\frac{1}{1+\epsilon-x}} \geq \sqrt{\frac{(2+\epsilon)(\epsilon+x)}{2}} \tag{5.17}$$

for each $0 \leq x \leq 1$ and $\epsilon > 0$ sufficiently small. On the other hand,

$$\sqrt{\frac{2}{b(b-1)}} \leq \sqrt{\frac{b(b-1)}{2}}\,, \tag{5.18}$$

for b sufficiently large. Using (5.14), (5.15), (5.16), (5.17) and (5.18) we conclude that there exist values of b for which the LHS of (5.13) becomes larger than, and values for which it becomes smaller than, the RHS of (5.13). Hence, there exist $b_0 > 1$ such that the LHS and the RHS of (5.13) are equal. Therefore, it is possible to choose a, b and B for which (5.11) and (5.12) are verified. For such choices, the 1-forms α_1, α_2 and α_3 do not have real periods and can be used to define a minimal surface $x\colon M \to R^3$. Since g and ω have poles at ∞, we easily conclude that M is complete in the induced metric.

The function g is meromorphic and can be extended to $\hat{M}$. Its image covers three times each point of $S^2(1)$. Hence, the total curvature of M is -12π. Finally, Jorge-Meeks formula (III 2.11) tell us that

$$6 = I - \chi(M),$$

where I is the multiplicity of the only end of M. Since $\chi(M) = -3$, then $I = 3$. This concludes the example.

6. A complete minimal surface of genus one with one end and total curvature -8π.

The next example to be described is also due to Chen and Gackstätter [1] and its construction follows the same method used in the previous section. For this reason, details are omitted. Let $\hat{M}$ be the Riemann surface where the function $w(z)$, given by

$$w^2(z) = z(z^2-a^2), \tag{6.1}$$

is well defined. We are going to assume that a is a positive real constant. The surface $\hat{M}$ is obtained by cutting the sphere $\mathbb{C} \cup \{\infty\}$ along two curves connecting $-a$ to 0 and a to ∞, and then by pasting two copies of this slit sphere along the slits, as was done in Section 5. The resulting surface $\hat{M}$ is a torus and hence, has genus one.

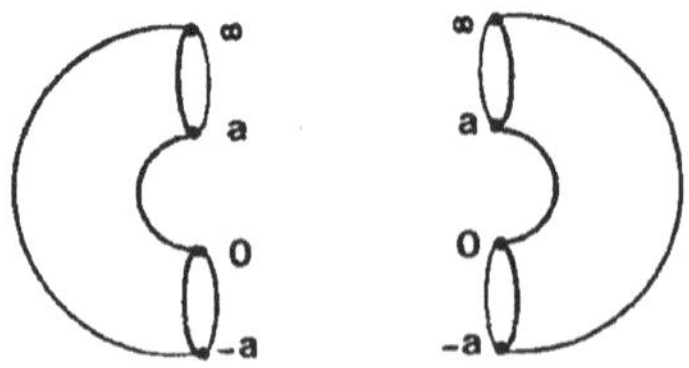

Fig. 18

Take $M = \hat{M} - \{\infty\}$, $g: M \to \mathbb{C} \cup \{\infty\}$ defined by

(6.2)
$$g(z) = B\, w(z)/z \ ,$$
$$\omega = z\, dz/w(z) \ ,$$

where B is a constant. It is easy to verify that, at $z = 0$ g has a pole of order one and ω has a zero of order two, and that $z = a$ and $z = -a$ are regular points of the form ω. From (6.2) we obtain

(6.3)
$$\alpha_1 = \frac{1}{2}\left(\frac{z}{w} - B^2 \frac{w}{z}\right) dz \ ,$$
$$\alpha_2 = \frac{i}{2}\left(\frac{z}{w} + B^2 \frac{w}{z}\right) dz \ ,$$
$$\alpha_3 = B\, dz \ ,$$

which we want not to have real periods. The existence of real periods must be searched among the cycles that generate the fundamental group of M. The fundamental group of the surface M is generated by two classes of homotopy. We may choose one curve in each one of them as indicated in the picture below, where the dark lines and the light lines stand for the trajectories on different copies of the slit

(6.3)

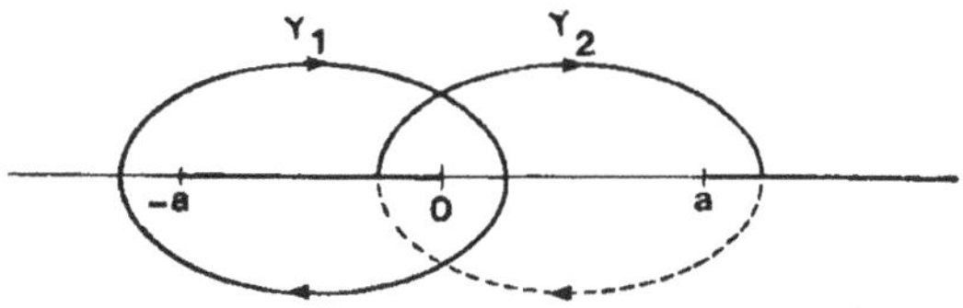

Fig. 19

plane used to construct $\hat{M}$. Since α_3 is exact, we have

$$(6.5) \qquad \int_{\gamma_k} \alpha_3 = 0, \qquad 1 \le k \le 2 .$$

Consider the positive real constants F and G defined by

$$(6.6) \qquad F = \int_0^a \frac{x\,dx}{\sqrt{x(a^2-x^2)}} \quad \text{and} \quad G = \int_0^a \frac{(a^2-x^2)\,dx}{\sqrt{x(a^2-x^2)}} .$$

The computation of the periods of α_j along γ_k is straightforward and the result is presented in the following table:

(6.7)

	$\gamma=\gamma_1$	$\gamma=\gamma_2$
$\int_\gamma \alpha_1$	$-F+B^2G$	$-i(F+B^2G)$
$\int_\gamma \alpha_2$	$-i(F+B^2G)$	$F-B^2G$

Thus, the nonexistence of real periods occurs when $B = \sqrt{F/G}$. For such a choice, the forms α_1, α_2 and α_3 can be used to define a minimal surface $x: M \to R^3$. Since g and ω have poles at ∞, we can conclude that M, in the induced metric, is complete.

The function g is meromorphic and can be extended to $\hat{M}$. Its image covers two times each point of $S^2(1)$. Hence, the total curvature of M is -8π. By Jorge-Meeks formula (III 2.11), we have

$$4 = I - \chi(M) ,$$

where I is the multiplicity of the only end of M. Since $\chi(M) = -1$, then $I = 3$. This concludes the example.

7. A complete minimal surface of genus one with three ends and total curvature -12π.

The following example was first described by Costa [1], who constructed it by making use of the classical Weierstrass $\wp$-function. However, it is possible to exhibit such an example as an application of the method used in the previous two sections. In what follows we are going to introduce this example by making both constructions.

First construction: Let $\hat{M}$ be the Riemann surface where the function $w(z)$, given by

$$w^2 = z(z^2-a^2) , \tag{7.1}$$

is well defined. We assume that a represents a positive real constant. As we have seen in the previous section, the surface $\hat{M}$ has genus one.

Take $M = \hat{M} - \{a,-a,\infty\}$, $g: M \to \mathbb{C} \cup \{\infty\}$ defined by

$$\begin{aligned} g(z) &= B/w , \\ \omega &= z\,dz/w , \end{aligned} \tag{7.2}$$

where B is a real constant. It is easy to verify that g has 3 poles of order one, precisely at $z = 0$, $z = a$ and $z = -a$, and has a zero of order three at $z = \infty$. On the other hand, ω has a double zero at $z = 0$ and a unique double pole at $z = \infty$.

From (7.2) we obtain

$$\begin{aligned} \alpha_1 &= \frac{1}{2}\left(\frac{z}{w} - \frac{B^2 z}{w^3}\right)dz , \\ \alpha_2 &= \frac{i}{2}\left(\frac{z}{w} + \frac{B^2 z}{w^3}\right)dz , \\ \alpha_3 &= \frac{Bz}{w^2}\,dz , \end{aligned} \tag{7.3}$$

which we want not to have real periods. The existence of real periods must be searched among the cycles that generate the fundamental group of M. These are the ones that generate the fundamental group of $\hat{M}$ and the ones around the poles of α_1, α_2 and α_3. The cycles may be chosen as indicated in the picture below, where the dark lines and the

(7.4) 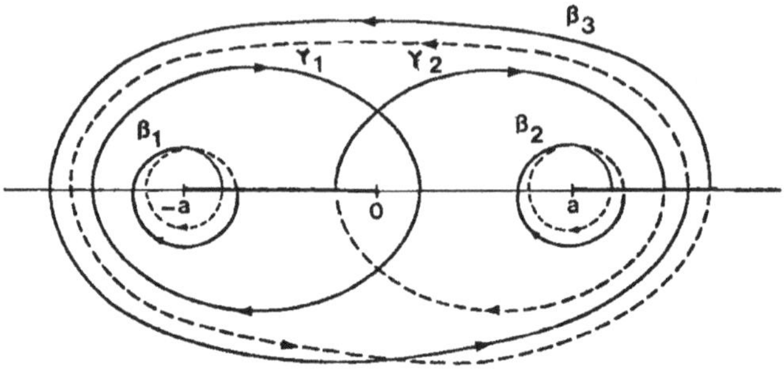

Fig. 20

light lines stand for the trajectories on different copies of the slit plane used to construct $\hat{M}$. The curves β_1, β_2 and β_3 make one turn around the points $-a$, a and ∞, respectively, and the curves γ_1 and γ_2 generate the homotopy of $\hat{M}$.

Observe that, on both expressions of α_1 and α_2, we have only odd powers of w. Since each β_k is the sum of two circles (one on each leaf) it is clear that

(7.5) $$\int_{\gamma_k} \alpha_1 = \int_{\gamma_k} \alpha_2 = 0, \qquad 1 \leq k \leq 3.$$

On the other hand, it is easy to conclude, by using residues, that

(7.6) $$\int_{\beta_k} \alpha_3 = \int_{\beta_k} \frac{B\, dz}{z^2-a^2} = \begin{cases} \pm \dfrac{2\pi B i}{2a}, & \text{if} \quad 1 \leq k \leq 2 \\ \\ 0 \quad , & \text{if} \quad k = 3 . \end{cases}$$

Consider the positive real constants

(7.7) $$F = \int_0^a \frac{x\,dx}{\sqrt{x(a^2-x^2)}} \quad \text{and} \quad G = \int_0^a \frac{dx}{(a^2-x^2)\sqrt{x(a^2-x^2)}} .$$

A simple computation of the periods of α_j along γ_1 and γ_2 furnishes data for the following table:

(7.8)

	$\gamma=\gamma_1$	$\gamma=\gamma_2$
$\int_\gamma \alpha_1$	$-F+B^2G$	$-i(F+B^2G)$
$\int_\gamma \alpha_2$	$-i(F+B^2G)$	$F-B^2G$
$\int_\gamma \alpha_3$	$-4\pi Bi$	$-4\pi Bi$

Thus, for the nonexistence of real periods, we must have $B = \sqrt{F/G}$. With this choice the forms α_1, α_2 and α_3 can be used to define a minimal surface $x: M \to R^3$. Since g has poles at $z = a$ and $z = -a$, and ω has a unique pole at $z = \infty$, we can conclude that M, in the induced metric, is complete.

The function g is meromorphic and extends to $\hat{M}$. Its image covers three times each point of $S^2(1)$. Thus, M has total curvature -12π. By Jorge-Meeks formula (III 2.11) we have

$$6 = \sum_{k=1}^{3} I_k - \chi(M) .$$

Since $\chi(M) = -3$, then $I_1 = I_2 = I_3 = 1$. Therefore, each end of M is embedded.

Recently, D. Hoffman and W. Meeks [1] proved that this example is embedded.

<u>Second construction</u>: Let $\hat{M} = \mathbb{C}/L$ represent the torus obtained as the quotient space of the complex plane $\mathbb{C}$ by the lattice $L = \{m+ni;\ m,m \in \mathbb{Z}\}$. Represent by π the canonical projection from $\mathbb{C}$ onto $\hat{M}$, and by $\wp : \mathbb{C} \to \mathbb{C} \cup \{\infty\}$ the Weierstrass function associated to the lattice L. The function $\wp$ is doubly periodic in the sense that $\wp(z+w) = \wp(w)$, for any $w \in L$ and any $z \in \mathbb{C}$. Hence, $\wp$ gives rise to a function $\hat{\wp}: \hat{M} \to \mathbb{C} \cup \{\infty\}$ defined by $\hat{\wp}\circ\pi = \wp$. The function $\wp$ is given by

$$(7.9)\qquad \wp(z) = \frac{1}{z^2} + \sum_{\substack{(m,n)\in\mathbb{Z}^2\\(m,n)\neq(0,0)}} \left[\frac{1}{(z-m-ni)^2} - \frac{1}{(m+ni)^2}\right] .$$

It satisfies the differential equation

$$(7.10)\qquad \wp'^2 = 4(\wp^2-a^2)\wp ,$$

where $a > 0$ and

$$(7.11)\qquad a = \wp(1/2) = -\wp(i/2) \text{ and } \wp((1+i)/2) = 0.$$

It is important to observe that $\wp'$ is also doubly periodic and gives rise to a new function $\hat{\wp}' : \hat{M} \to \mathbb{C} \cup \{\infty\}$. If we write $w = \wp'$ and $z = \wp$, then we obtain from (7.10)

$$(7.12)\qquad w^2 = 4z(z^2-a^2) .$$

This is, up to constant factor, the equation (7.1) used in the first construction. The table below summarizes the relevant informations about $\wp$ and $\wp'$ at the points 0, 1/2, i/2 and (1+i)/2; the

(7.12)

z	0	1/2	i/2	(1+i)/2
$\wp$	2	a	-a	0^2
$\wp'$	3	0^1	0^1	0^1

superscripts indicate the orders of the poles, or the order of the zeroes, according to the case. It will be relevant in this construction to know the values of the integrals of $\wp$ along the cycles that generate the fundamental group of the torus $\hat{M}$. One can show that if $\beta(t) = t+ib$ and $\gamma(t) = b+it$, with $b - \frac{1}{2} \notin \mathbb{Z}$ and $0 \leq t \leq 1$, then

$$(7.13)\qquad \begin{aligned} \int_\beta \wp(z)dz &= \int_\beta \wp(z - \tfrac{1}{2})dz = \int_\beta \wp(z - \tfrac{i}{2})dz = -\pi \\ \int_\gamma \wp(z)dz &= \int_\gamma \wp(z - \tfrac{1}{2})dz = \int_\gamma \wp(z - \tfrac{i}{2})dz = i\pi . \end{aligned}$$

The so called "addition theorem" for the function $\wp$ (Siegel [1] pp. 80-82) states that

$$(7.14)\qquad \wp(z_1+z_2) = \frac{1}{4}\left[\frac{\wp'(z_1)-\wp'(z_2)}{\wp(z_1)-\wp(z_2)}\right]^2 - \wp(z_1) - \wp(z_2) .$$

By using (7.14) and (7.10) one can conclude that

$$(7.15)\qquad \begin{aligned} \wp(z - \tfrac{1}{2}) - a &= 2a^2/(\wp(z)-a), \\ \wp(z - \tfrac{i}{2}) + a &= 2a^2/(\wp(z)+a), \\ \wp(z - \tfrac{1+i}{2}) &= -a^2/\wp(z) . \end{aligned}$$

The last fact that will be necessary to know about the function $\wp$ for the construction of the example, is its Laurent development in power series around $z = 0$, which is given by

$$(7.16)\qquad \wp(z) = z^{-2} + \sum_{n=1}^{\infty} b_n z^{2n} .$$

All properties of the function $\wp$ mentioned above are classical, and can be found in Siegel [1], Neville [1] or Farkas [1].

Set $p_1 = \pi(0)$, $p_2 = \pi(\frac{1}{2})$, $p_3 = \pi(\frac{i}{2})$ and $p_4 = \pi(\frac{1+i}{2})$. Take $M = \hat{M} - \{p_1, p_2, p_3\}$, $g\colon M \to \mathbb{C} \cup \{\infty\}$ defined by

$$(7.17)\qquad \begin{aligned} g(\pi(z)) &= A/\wp'(z) , \\ \omega_{\pi(z)} &= \wp(z)dz , \end{aligned}$$

where A is a real constant. From (7.12) it follows that the zeroes and poles of g and ω occur at p_j, $1 \le j \le 4$, as indicated in the following table:

(7.18)

	p_1	p_2	p_3	p_4
g	0^3	∞^1	∞^1	∞^1
ω	∞^2			0^2

.

By using (7.2) and (7.10) we obtain

$$(7.19)\qquad \begin{aligned} \alpha_1 &= \frac{1}{2}\left(\wp - \frac{A^2}{4(\wp^2-a^2)}\right)dz , \\ \alpha_2 &= \frac{i}{2}\left(\wp + \frac{A^2}{4(\wp^2-a^2)}\right)dz , \\ \alpha_3 &= \frac{A\wp'}{4(\wp^2-a^2)}\, dz \end{aligned}$$

which must not have real periods in M. For this to happen it suffices that the following conditions hold:

a) for each $1 \le k \le 3$, $1 \le j \le 3$, $\operatorname{Res}(\alpha_k)_{p_j}$ is a real number;

b) if $\hat{\beta}$ and $\hat{\gamma}$ are generators for the fundamental group of $\hat{M}$, then $\int_{\hat{\beta}} \alpha_k$ and $\int_{\hat{\gamma}} \alpha_k$ are purely imaginary, $1 \le k \le 3$.

The verification of these conditions are simplified after we rewrite each α_k in a proper form. For this, observe that

$$\frac{1}{\wp^2 - a^2} = \frac{1}{2a}\left(\frac{1}{\wp - a} - \frac{1}{\wp + a}\right)$$

and make use of (7.15) to obtain

$$(7.20) \qquad \frac{1}{\wp^2(z) - a^2} = \frac{1}{4a^3}\left(\wp\left(z - \tfrac{1}{2}\right) - \wp\left(z - \tfrac{i}{2}\right) - 2a\right).$$

Thus,

$$\alpha_1 = \frac{1}{2}\left[\wp(z) - \frac{A^2}{16a^3}\left(\wp\left(z - \tfrac{1}{2}\right) - \wp\left(z - \tfrac{i}{2}\right) - 2a\right)\right]dz ,$$

$$(7.21) \qquad \alpha_2 = \frac{i}{2}\left[\wp(z) + \frac{A^2}{16a^3}\left(\wp\left(z - \tfrac{1}{2}\right) - \wp\left(z - \tfrac{i}{2}\right) - 2a\right)\right]dz ,$$

$$\alpha_3 = \frac{1}{8a}\left(\frac{\wp'(z)}{\wp(z) - a} - \frac{\wp'(z)}{\wp(z) + a}\right)dz .$$

Since $\wp$ and $\wp'$ are doubly periodic, it suffices to make the computations of the residues of α_k at the points $z = \frac{1}{2}$, $z = \frac{i}{2}$ and $z = 0$, which are in the inverse image of the points p_2, p_3 and p_1, respectively. Since the only pole of $\wp$ is $z = 0$, where its residue is zero (according to (7.16)),

$$(7.22) \qquad \mathrm{Res}(\alpha_1)_z = \mathrm{Res}(\alpha_2)_z = 0, \quad \text{for any } z \in \mathbb{C}.$$

Let γ_ε be a circle of sufficiently small radius $\varepsilon > 0$, around $z = 1/2$. Since $\wp$ is holomorphic in a neighborhood of $z = 1/2$, we may take $w = \wp(z)$ along γ_ε and then

$$\lim_{\varepsilon \to 0} \int_{\gamma_\varepsilon} \alpha_3 = \frac{A}{8a} \lim_{\varepsilon \to 0} \int_{\wp(\gamma_\varepsilon)} \frac{dw}{w - a} = \frac{A}{8a}\, 2\pi i n ,$$

where n is an integer greater than or equal to two. Thus,

(7.23) $$\operatorname{Res}(\alpha_3)_{z=\frac{1}{2}} = \frac{nA}{8a} .$$

This argument may be repeated when $z = i/2$, and we obtain

(7.24) $$\operatorname{Res}(\alpha_3)_{z=\frac{i}{2}} = -\frac{mA}{8a} ,$$

where m is an integer. By using (7.16) we obtain

$$\frac{\wp'(z)}{\wp(z)\pm a} = \frac{-\frac{2}{z} + \sum_{n=1}^{\infty} 2nb_n z^{2n+1}}{1\pm az^2 + \sum_{n=1}^{\infty} b_n z^{2n+2}} = \frac{f(z)}{2} + h(z) ,$$

with f and h holomorphic and $f(0) = -2$. It follows that

(7.25) $$\operatorname{Res}(\alpha_3)_{z=0} = 0 .$$

Let β and γ be the curves used in (7.13). Observe that $\hat{\beta} = \pi\circ\beta$ and $\hat{\gamma} = \pi\circ\gamma$ are generators for the fundamental group of $\hat{M}$. Using (7.13) we now obtain

$$\int_{\hat{\beta}} \alpha_1 = \frac{1}{2} [-\pi - \frac{A^2}{16a^3} (-\pi+\pi-2a) = \frac{1}{2} (-\pi + \frac{A^2}{8a^2}) ,$$

$$\int_{\hat{\gamma}} \alpha_1 = \frac{1}{2} [i\pi - \frac{A^2}{16a^3} (i\pi-i\pi-2ai) = \frac{i}{2} (\pi + \frac{A^2}{8a^2}) .$$

In an analogous way,

$$\int_{\hat{\beta}} \alpha_2 = -\frac{i}{2} (\pi + \frac{A^2}{8a^2}) ,$$

(7.27) $$\int_{\hat{\gamma}} \alpha_2 = -\frac{1}{2} (\pi - \frac{A^2}{8a^2}) .$$

On the other hand, if $w_o \neq 0$ is a real number, then

$$\int_{\hat{\beta}} \frac{\wp'(z)dz}{\wp(z)-w_o} = \log(\wp(bi+1)-w_o) - \log(\wp(bi)-w_o) .$$

Since $\wp$ is periodic, $\wp(bi+1) = \wp(bi)$ and this integral is zero. Therefore,

$$\int_{\hat{\beta}} \alpha_3 = 0 . \tag{7.28}$$

By analogous reason,

$$\int_{\hat{\gamma}} \alpha_3 = 0 . \tag{7.29}$$

The results obtained from (7.22) to (7.29) show that, by choosing A such that $A^2 = 8\dot{a}^2\pi$, the forms α_k have no real periods. Thus, they can be used to define a minimal immersion $x\colon M \to R^3$. Since p_1, p_2, and p_3 are poles of g or ω, then M, in the induced metric, is complete. It is easy to verify that g covers each point of $\mathbb{C} \cup \{\infty\}$ three times; hence, the total curvature of M is -12π. Making use of the Jorge-Meeks formula, as we did before, we can also conclude that each one of the three ends of M is embedded.

Observe that, at p_1, g has one zero of order three while ω has one pole of order two. Thus, $\alpha_3 = g\omega$ is a well defined form at p_1 and the coordinate $x_3 = \mathrm{Re}\int^w \alpha_3$ of the immersion x has a finite limit when w converges to p_1. This means that the end of $x(M)$, corresponding to $w = p_1$, approaches asymptotically a plane. Since $g(p_2) = g(p_3) = \infty$ and $g(p_1) = 0$, then the three ends are parallel in R^3.

The following is the picture of a compact piece of Costa's example included in Hoffman-Meeks [1]; it is reproduced here with the permission of the authors.

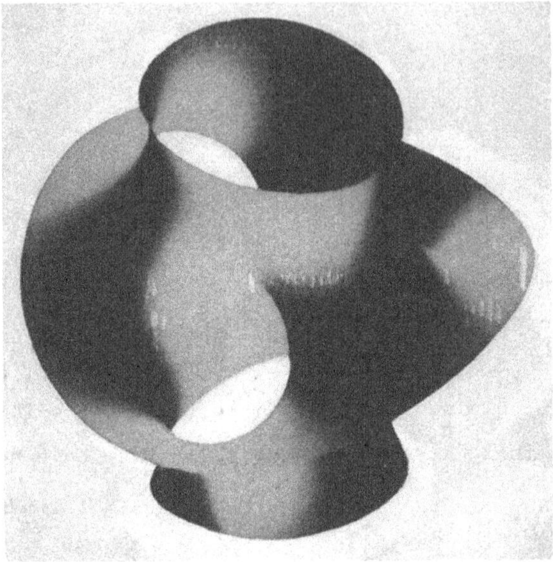

Fig. 21

Costa's surface*

8. A complete minimal surface of genus one with two ends and total curvature -20π.

In this section we introduce another example of Costa. This is a complete minimal surface in R^3 with total curvature -20π, of genus one with two ends. We are going to use the Weierstrass $\wp$-function associated to the lattice $L = \{m+in;\ m,n \in \mathbb{Z}\}$, whose properties has been pointed out in the previous section.

Set $\hat{M} = \mathbb{C}/L$, $\pi: \mathbb{C} \to \hat{M}$ the canonical projection, $q_1 = \pi(0)$,

*The computer graphic of this picture was devised by J.T. Hoffman.

$q_2 = \pi(\frac{1+i}{2})$, $q_3 = \pi(\frac{1}{2})$, $q_4 = \pi(\frac{i}{2})$, $M = \hat{M} - \{q_1, q_2\}$, $g: M \to \mathbb{C} \cup \{\infty\}$ defined by

$$g(\pi(z)) = A/\wp(z)\wp'(z),$$
(8.1)
$$\omega_{\pi(z)} = (\wp'(z))^2 dz ,$$

where A is a real constant. From (7.12) it follows that the zeroes and poles of g and ω occur at q_1, q_2, q_3 and q_4, as indicated in the following table:

(8.2)

	q_1	q_2	q_3	q_4
g	0^5	∞^3	∞^1	∞^1
ω	∞^6	0^2	0^2	0^2

By using (8.1) above and (7.10) of the previous section, we obtain

$$\alpha_1 = \frac{1}{2}\,(4\wp^3 - 4a^2\wp - \frac{A^2}{2})dz ,$$
(8.3)
$$\alpha_2 = \frac{i}{2}\,(4\wp^3 - 4a^2\wp + \frac{A^2}{2})dz ,$$
$$\alpha_3 = \frac{A\wp'}{\wp}\,dz ,$$

which we want to have no real periods in M. For this to happen it suffices that the following conditions hold in $\hat{M}$:

a) for each $1 \le k \le 3$, $1 \le j \le 2$, $\mathrm{Res}(\alpha_k)_{q_j}$ is a real number;

b) if $\hat{\beta}$ and $\hat{\gamma}$ are cycles that generate the fundamental group of $\hat{M}$, then $\int_{\hat{\beta}} \alpha_k$ and $\int_{\hat{\gamma}} \alpha_k$ are purely imaginary, $1 \le k \le 3$.

The verification of these conditions are simplified after we rewrite each α_k in a proper form. For this, we start by taking derivatives on both sides of (7.10) to obtain

$$(8.4) \qquad \rho'' = 6\rho^2 - 2a^2 .$$

Making use of (7.10) and (8.4) above, it follows that

$$(8.5) \qquad (\rho\rho')' = 10\ {}^3 - 6a^2\rho .$$

Thus,

$$(8.6) \qquad \rho^3 = \frac{1}{10}(\rho\rho')' + \frac{6}{10}a^2\rho .$$

Substitution of this into the expression of α_1 in (8.3) yields

$$(8.7) \qquad \alpha_1 = \frac{1}{5}((\rho\rho')' - 4a^2\rho)dz - \frac{A^2}{2\rho^2}dz .$$

Making use of (7.15) to replace the value of $1/\rho^2$, we obtain

$$(8.8) \qquad \alpha_1 = \frac{1}{5}((\rho\rho')' - 4a^2\rho)dz - \frac{A^2}{2a^4}\rho^2(z - \frac{1+i}{2})dz .$$

It follows now from (8.4) above that

$$(8.9) \qquad \alpha_1 = \frac{1}{5}((\rho\rho')' - 4a^2\rho)dz - \frac{A^2}{12a^4}(\rho''(z - \frac{1+i}{2}) + 2a^2)dz .$$

In a similar way, one shows that

$$(8.10) \qquad \alpha_2 = \frac{i}{2}((\rho\rho')' - 4a^2\rho)dz + \frac{iA^2}{12a^4}(\rho''(z - \frac{1+i}{2}) + 2a^2)dz .$$

It is not necessary to make any change in the expression of α_3.

It is now a straighforward computation, making use of (7.18) and (7.16), to show that

$$(8.11) \qquad \operatorname{Res}(\alpha_j)_{q_k} = 0, \qquad 1 \le j \le 2, \quad 1 \le k \le 2.$$

By using (7.16) in the expression of α_3, we conclude that

$\alpha_3 = A(-\frac{2}{z} + f(z))dz$ around $z = 0$, where f is holomorphic. Therefore,

(8.12)
$$\operatorname{Res}(\alpha_3)_{q_1} = -2A.$$

To compute the residue of α_3 at q_2, we first use (7.15) to rewrite α_3 as

(8.13)
$$\alpha_3 = -\frac{A}{a^2} \wp'(z)\, \wp(z - \frac{i+1}{2})dz .$$

By making use of the Laurent power series development of $\wp$ around $z = 0$, observing that the first term of the power series development of $\wp'(z)$ around $z = \frac{1+i}{2}$ is $\wp''(\frac{1+i}{2})(z - \frac{1+i}{2})$, and observing that (8.4) above yields $\wp''(\frac{1+i}{2}) = -2a^2$, we obtain

(8.14)
$$\alpha_3 = \left[\frac{2A}{z-\frac{1+i}{2}} + h(z)\right]dz ,$$

where $h(z)$ is a holomorphic function defined in a neighborhood of $z = (1+i)/2$. Thus,

(8.15)
$$\operatorname{Res}(\alpha_3)_{p_2} = 2A .$$

It is sufficient to compute the periods of α_k along the curves $\hat{\beta} = \pi(\beta)$ and $\hat{\gamma} = \pi(\gamma)$, where β and γ are the curves used in (7.13). Since $\wp$ and $\wp'$ are doubly periodic with respect to the same lattice, then

(8.16)
$$\begin{aligned}
\int_{\hat{\beta}} \alpha_1\, dz &= \frac{4}{5} a^2\pi - \frac{A^2}{6a^2} ,\\
\int_{\hat{\gamma}} \alpha_1\, dz &= -\frac{4}{5} a^2\pi i - \frac{A^2}{6a^2} i ,\\
\int_{\hat{\beta}} \alpha_2\, dz &= \frac{4}{5} a^2\pi i + \frac{A^2}{6a^2} i ,\\
\int_{\hat{\gamma}} \alpha_2\, dz &= \frac{4}{5} a^2\pi - \frac{A^2}{6a^2} .
\end{aligned}$$

Finally,

$$(8.17)\qquad \int_{\hat{\beta}} \alpha_3 = A\int_{\beta} \frac{\wp'(z)}{\wp(z)}\,dz = A(\log \wp(\beta(1)) - \log \wp(\beta(0)) = 0\ ,$$

where the last equality comes from the periodicity of $\wp$ and the fact that $\beta(1) - \beta(0) = 1$. In an analogous way, we obtain

$$(8.18)\qquad \int_{\hat{\gamma}} \alpha_3 = 0\ .$$

Thus, if we take $A^2 = \frac{24}{5} a^4 \pi$, the forms α_k, $1 \leq k \leq 3$, do not have real periods. Therefore, they can be used to define a minimal immersion $x: M \to R^3$. Since ω has one pole in q_1 and $g^2\omega$ has a pole at q_2, it follows that M, in the induced metric, is complete. Since $g: \hat{M} \to \mathbb{C} \cup \{\infty\}$ is meromorphic and g assume the value zero with order five, then g covers each point of $\mathbb{C} \cup \{\infty\}$ five times. Therefore the total curvature of M is -20π. This concludes the example.

9. A complete minimal surface between two parallel planes.

This example is due to Jorge and Xavier [1]. Consider $M = D = \{z \in \mathbb{C};\ |z| < 1\}$, $g: D \to \mathbb{C} \cup \{\infty\}$ defined by

$$(9.1)\qquad \begin{aligned} g(z) &= e^{f(z)}\ , \\ \omega &= e^{-f(z)}dz\ , \end{aligned}$$

where f is a function whose construction is described below.

Let D_n be a sequence of concentric disks such that the closure of D_n is contained in the interior of D_{n+1} and $\bigcup D_n = D$. Let K_n be the compact set obtained by taking an annulus contained in $D_{n+1} - \bar{D}_n$ and deleting from it a small piece (as indicated in the picture below) which contains the intersection of D_{n+1} with the positive x axis, if n is even, or with the negative x axis, if n is odd. We want $D-K_n$ to be connected.

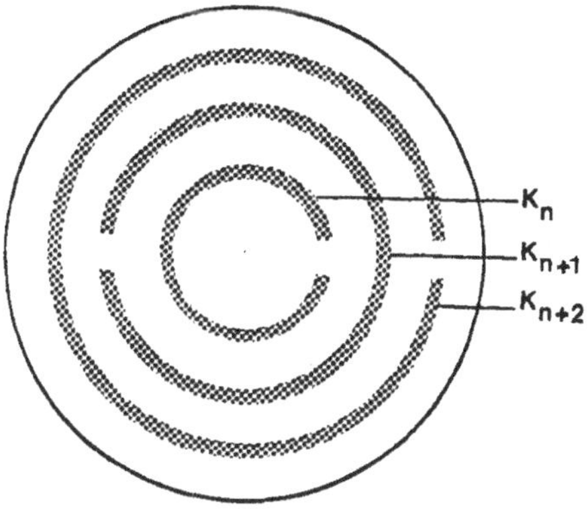

Fig. 22

Take open disjoint sets U_n such that $U_n \supset K_n$ and choose real constants c_n, $1 \leq n < \infty$; now define $h\colon \bigcup U_n \to \mathbb{C}$ by $h(z) = c_n$ for $z \in U_n$.

It is a consequence of Runge's theorem (cfr. Rudin [1]) that there exists a holomorphic function $f\colon D \to \mathbb{C}$ such that, for each n,

$$|f(z)-c_n| < 1 \quad \text{in} \quad K_n .$$

That is the function f that we are going to use in (9.1) to define g and ω. With this choice, we will show that g and ω give rise, through the Weierstrass formulas, to a minimal immersion $x\colon D \to R^3$ which induces on D a complete metric and whose image lies between two parallel planes.

First observe that $\alpha_3 = dz$ and hence, $x_3 = \mathrm{Re}(z)$. Since $|z| < 1$, $-1 < \mathrm{Re}(z) < 1$. Therefore, $x(D)$ is contained in the region bounded by the planes $x_3 = 1$ and $x_3 = -1$.

To show that M is complete in the induced metric, it suffices to show that any divergent path in D has infinite length in the metric

$$ds^2 = \frac{1}{4}\left(\frac{1}{|g|} + |g|\right)^2 |dz|^2 .$$

Let $\gamma: [a,b) \to D$ be a divergent path in D parametrized by the Euclidean arc length in the disk. Since $\frac{1}{2}\left(\frac{1}{|g|} + |g|\right) \geq 1$ then, if γ has infinite Euclidean length $(b = \infty)$,

$$L(\gamma) = \int_a^\infty \frac{1}{2}\left(\frac{1}{|g|} + |g|\right)dt \geq \int_a^\infty dt = \infty.$$

If γ has finite Euclidean length, then γ cuts all but a finite number of K_n with n even, or it cuts all but a finite number of K_n with n odd. Suppose the first case occurs. Set

$$J_n = \{t \in [a,b);\ \gamma(t) \in K_n\} .$$

Then we have

$$L(\gamma) = \int_a^b \frac{1}{2}\left(\frac{1}{|g|} + |g|\right)dt \geq \int_a^b |g|dt \geq$$

$$\geq \sum_{\substack{n>N \\ n \text{ even}}} \int_{J_n} |g|dt .$$

Observe that, on J_n, we have

$$|g| = |e^f| = |e^{f-c_n} e^{c_n}| \geq e^{-1} e^{c_n} = e^{c_n - 1} .$$

Thus,

$$L(\gamma) \geq \sum_{\substack{n>N \\ n \text{ even}}} \int_{J_n} e^{c_n - 1}\, dt \geq \sum_{\substack{n>N \\ n \text{ even}}} r_n\, e^{c_n - 1},$$

where r_n is the width of K_n. Hence, it is enough to choose each c_n in such way that $e^{c_n - 1}$ grows sufficiently fast. One can, for example, take $c_n = -\log r_n$. In this case, $r_n\, e^{c_n - 1} = e^{-1}$ and the series

$$\sum_{\substack{n>N \\ n \text{ even}}} r_n\, e^{c_n - 1} \quad \text{diverges.}$$

The other case is treated similarly, and this completes the example.

CHAPTER V

NONEXISTENCE OF CERTAIN MINIMAL SURFACES

We have exhibited, along this notes, a great variety of examples of complete minimal surfaces in R^3. We are going to dedicate this chapter to the proof of two theorems concerned with the nonexistence of certain types of minimal surfaces. The first, due to Jorge-Meeks [1], is the following:

(1.1) THEOREM. <u>The only complete finite total curvature minimal embeddings of</u> $S^2 - \{p_1,\dots,p_k\}$ <u>in</u> R^3, <u>for</u> $1 \le k \le 5$, <u>are the plane</u> $(k=1)$ <u>and the catenoid</u> $(k=2)$. <u>The cases</u> $k = 3,4$ <u>or</u> 5 <u>do not occur.</u>

Proof: Let M be minimally embedded in R^3 and conformally diffeomorphic to S^2 minus k points. By Theorem (III 2.10) the total curvature of M is given by

$$C(M) = -4\pi(k-1). \tag{1.2}$$

Since the Gauss mapping g is a meromorphic function on S^2, then it is given by a quotient of two polynomials

$$g = \frac{Q}{P} \tag{1.3}$$

and

$$k-1 = \max\{\deg P, \deg Q\}. \tag{1.4}$$

The fact that g is meromorphic means, geometrically, that the Gauss mapping $N: S^2 - \{p_1,\dots,p_k\} \to S^2(1)$ extends to S^2. It is now a direct consequence of (III 2.10) that $N(p_1),\dots,N(p_k)$ must be parallel; otherwise M would not be embedded. After a change of coordinates in R^3 and M, we may assume that

(1.5) $$g(\{p_1,\dots,p_k\}) = \{1,-1\}$$

and that $\deg Q < \deg P$. Let a_j, $1 \leq j \leq t_0$, be the elements of $\{p_1,\dots,p_k\}$ such that $g(a_j) = 1$, and b_m, $1 \leq m \leq t_1$, be the elements of $\{p_1,\dots,p_k\}$ such that $g(b_m) = -1$. Then,

(1.6) $$\deg(P) = k-1 \quad \text{and} \quad k = t_0+t_1 .$$

Let Y_r, $r > 0$, be the intersection of M with a sphere of radius r centered at the origin, and set $X_r = \frac{1}{r} Y_r$. By Theorem (III 2.10), for large r, $X_r = \{Y_1^r,\dots,Y_k^r\}$ consists of k closed curves embedded in $S^2(1)$ (which converge to a unique great circle in $S^2(1)$ perpendicular to $N(p_j)$). Consequently, if C_r is a solid cylinder of radius r having as axis the straight line generated by $(1,0,0)$, then M divides C_r into two connected open sets. Furthermore, $M \cap \partial C_r$ is the union of k disjoint closed curves. Since M is oriented, the consecutive curves of $M \cap \partial C_r$ have opposite orientations. Consequently,

(1.7) $$t_0 = t_1 \quad \text{if } k \text{ is even}, \quad \text{and}$$
$$|t_0-t_1| = 1 \quad \text{if } k \text{ is odd}.$$

In the last case, (exchanging z by $-z$, if necessary) we may always assume that

(1.8) $$t_0+1 = t_1 , \quad \text{when } k \text{ is odd}.$$

The function f in the Weierstrass representation of M must have poles at a_j and b_m, which must have order two according to (III 2.4), and zeroes at the points where g has poles. Furthermore, the order of such zeroes must be two times the order of the corresponding poles. Thus,

$$f = P^2 / \prod_{j=1}^{t_o} (z-a_j)^2 \prod_{m=1}^{t_1} (z-b_m)^2 . \tag{1.9}$$

From this, we obtain

$$\begin{aligned} \alpha_1 &= (P^2-Q^2)dz/2\Pi(z-a_j)^2\Pi(z-b_m)^2 , \\ \alpha_2 &= i(P^2+Q^2)dz/2\Pi(z-a_j)^2\Pi(z-b_m)^2 , \\ \alpha_3 &= PQ\ dz/\Pi(z-a_j)^2\Pi(z-b_m)^2 . \end{aligned} \tag{1.10}$$

From (1.3) above it follows that $P(a_j) = Q(a_j)$ and $P(b_m) = -Q(b_m)$. Thus,

$$\begin{aligned} P-Q &= G\Pi(z-a_j)^{m_j} , \\ P+Q &= H\Pi(z-b_m)^{n_m} , \end{aligned} \tag{1.11}$$

where $m_j \geq 1$, $n_m \geq 1$, $G(a_j) \neq 0$, $G(b_m) \neq 0$, $H(a_j) \neq 0$ and $H(b_m) \neq 0$, for $1 \leq j \leq t_o$, $1 \leq m \leq t_1$. Since the degree of g is $k-1$, we have that $P-Q$ and $P+Q$ have $k-1$ zeroes. Therefore,

$$\deg(G) + \Sigma m_j = \deg(H) + \Sigma n_m = k-1. \tag{1.12}$$

Substitution of (1.11) into (1.10) gives

$$\alpha_1 = \frac{1}{2}\, GH\prod(z-a_j)^{m_j-2}\prod(z-b_m)^{n_m-2}\, dz,$$

(1.13)
$$\alpha_2 = \frac{i}{4}\left[\frac{G^2\prod(z-a_j)^{2m_j-2}}{\prod(z-b_m)^2} + \frac{H^2\prod(z-b_m)^{2n_m-2}}{\prod(z-a_j)^2}\right]dz\ ,$$

$$\alpha_3 = -\frac{1}{4}\left[\frac{G^2\prod(z-a_j)^{2m_j-2}}{\prod(z-b_m)^2} + \frac{H^2\prod(z-b_m)^{2n_m-2}}{\prod(z-a_j)^2}\right]dz\ .$$

Since each of such a form does not have real periods, it follows that

(1.14)
$$\frac{G^2\prod(z-a_j)^{2m_j-2}}{\prod(z-b_m)^2}\,dz \quad \text{and} \quad \frac{H^2\prod(z-b_m)^{2n_m-2}}{\prod(z-a_j)^2}\,dz$$

do not have real periods. We will need the following

(1.15) LEMMA. *If* $\prod_{j=1}^{n} (z-c_j)^{m_j}dz \Big/ \prod_{m=1}^{p} (z-d_m)^2$, *with* $c_j \neq d_m$ *for each* j *and* m *does not have periods, then*

$$\sum_{j=1}^{n} \frac{m_j}{d_i-c_j} = \sum_{\substack{m=1\\ m\neq i}}^{p} \frac{2}{d_i-d_m}\ ,$$

for each $i = 1,2,\ldots,n.$

Proof of the lemma: If the given forms have no periods, then they are exact and hence, for each $i = 1,\ldots,n,$ we have

$$0 = \frac{d}{dz}\left[\frac{\prod_{j=1}^{n} (z-c_j)^{m_j}}{\prod_{\substack{m=1\\ m\neq 1}}^{p} (z-d_m)^2}\right]_{z=d_i} =$$

$$= \frac{\prod_{j=1}^{n} (d_i - c_j)^{m_j}}{\prod_{\substack{m=1\\m\neq i}}^{p} (d_i - d_m)^2} \left(\sum_{j=1}^{n} \frac{m_j}{d_i - c_j} - \sum_{\substack{m=1\\m\neq i}}^{p} \frac{2}{d_i - d_m} \right) .$$

This proves the lemma.

Let us return to the proof of the theorem. Observe that

$$G = c_0 \prod (z - A_j)^{\delta_j} \quad \text{and} \quad H = c_1 \prod (z - B_j)^{\epsilon_j} .$$

A straighforward computation of the periods of (1.14) using the above lemma gives

$$(1.16) \qquad \Sigma \frac{\delta_j}{b_r - A_j} + \sum_{j=1}^{t_o} \frac{m_j - 1}{b_r - a_j} = \sum_{\substack{m=1\\m\neq r}}^{t_1} \frac{1}{b_r - b_m} , \qquad r = 1, \ldots, t_1 ,$$

$$(1.17) \qquad \Sigma \frac{\delta_j}{a_s - B_j} + \sum_{m=1}^{t_1} \frac{n_m - 1}{a_s - b_m} = \sum_{\substack{j=1\\j\neq s}}^{t_o} \frac{1}{a_s - a_j} , \qquad s = 1, \ldots, t_o .$$

If $k \in \{3,4,5\}$, then, by (1.7) and (1.8), we have that t_o or t_1 is equal to 2. If $t_o = 2$ then, from (1.12) we obtain

$$(1.18) \qquad \Sigma \epsilon_j + \sum_{j=1}^{t_1} (n_j - 1) = k - 1 - t_1 = t_o - 1 = 1.$$

Then, one of the two assertions hold:

(i) $\epsilon_{j_o} = 1$ and $\epsilon_j = 0$ for all $j \neq j_o$, and $n_j = 1$ for all j;

(ii) $\eta_{j_o} = 2$ and $n_j = 1$ for all $j \neq j_o$, and $\epsilon_j = 0$ for all j.

Thus, the equations (1.17) above are of the type

$$\frac{1}{a_1-B} = \frac{1}{a_1-a_2}, \qquad \frac{1}{a_2-B} = \frac{1}{a_2-a_1},$$

or

$$\frac{1}{a_1-b_1} = \frac{1}{a_1-a_2}, \qquad \frac{1}{a_2-b_1} = \frac{1}{a_2-a_1}.$$

These equalities lead us to a contradiction. Thus, $t_o \neq 2$. Analogously, we have $t_1 \neq 2$. Therefore, $t_1 = 1$ and $t_o = 0$ or 1, and so $k \notin \{3,4,5\}$. If $k = 1,2$, then M is a plane $(k=1)$ or the catenoid $(k=2)$, as we have seen in (III 2.23).

The second theorem that we are going to prove is due to Meeks [4] and is stated as follows:

(1.19) THEOREM. <u>If M is diffeomorphic to a projective plane minus two points, it does not exist a complete minimal immersion</u> $x: M \to R^3$ <u>with total curvature</u> -6π.

<u>Proof:</u> Set $M = \mathbb{P}^2 - \{p_o, p_1\}$. Let $\pi: \tilde{M} = S^2(1) - \{q_1, -q_1, q_2, -q_2\} \to M$ be its oriented two-sheeted covering. If $x: M \to R^3$ is a complete minimal immersion, so does the map $\tilde{x} = x \circ \pi$. Let $g: S^2(1) \to S^2(1)$ be its Gauss mapping extended to $S^2(1)$. We may choose suitable coordinates in $\tilde{M}$ such that two of its four ends correspond to $z = 0$ and $z = a$, where a is positive real number. Furthermore, we may assume that the order of branching of g at $z = 0$ is greater than or equal to its order of branching at $z = a$. The other two ends of $\tilde{M}$ will be at $z = \infty$ and $z = -1/a$. This is a consequence of the next lemma.

Observe that the transformation $I: \mathbb{C} \cup \{\infty\} \to \mathbb{C} \cup \{\infty\}$, given by $I(z) = -1/\bar{z}$, corresponds to the antipodal mapping of $S^2(1)$. A necessary and sufficient condition for the immersion $\tilde{x}$ to be factored as $\tilde{x} = x \circ \pi$, where $x: U \subset \mathbb{P} \to R^3$, is that $\tilde{x}(I(z)) = \tilde{x}(z)$, for each z in $\tilde{U} = \pi^{-1}(U)$. If $\tilde{x}$ is a minimal immersion, this last condition is equivalent to two conditions about f and g, as stablished in the lemma below.

(1.20) LEMMA. *Let* $\tilde{x}: \tilde{U} \subset \mathbb{C} \cup \{\infty\} \to R^3$ *be a minimal immersion and* f *and* g *be the functions associated to* x *by the Weierstrass representation. Then,* $\tilde{x}(I(z)) = \tilde{x}(z)$ *for all* z *in* $\tilde{U}$ *if and only if the following occurs:*

a) $g(I(z)) = I(g(z))$ *for each* z *in* U, *and*

b) $f(z) = -\overline{f(I(z))} / (zg(z))^2$.

Proof of the lemma: If $\tilde{x}(I(z)) = \tilde{x}(z)$ and $w = I(z)$, then

$$\phi(z) = \frac{\partial \tilde{x}}{\partial z} = \frac{\partial}{\partial z} \tilde{x}(I(z)) = \frac{\partial \tilde{x}}{\partial \bar{w}} \frac{d\bar{I}}{dz} = \frac{1}{z^2} \overline{\phi(I(z))}.$$

Using (II 1.18), we obtain

$$f(I(z)) = \phi_1(I(z)) - i\phi_2(I(z)) = \bar{z}^2 \overline{(\phi_1(z) + i\phi_2(z))} =$$

$$= \frac{\bar{z}^2 \overline{(\phi_1^2(z) + \phi_2^2(z))}}{\overline{\phi_1(z) - i\phi_2(z)}} = \frac{-\bar{z}^2 \, \bar{\phi}_3^2(z)}{\overline{(\phi_1(z) - i\phi_2(z))^2}} \overline{(\phi_1(z) - i\phi_2(z))} =$$

$$= -\bar{z}^2 \, \overline{g(z)}^2 \overline{f(z)}$$

and

$$g(I(z)) = \frac{\phi_3(I(z))}{\phi_1(I(z)) - i\phi_2(I(z))} = \frac{\bar{z}^2 \overline{\phi_3(z)}}{-\bar{z}^2 \overline{g(z)}^2 \overline{(\phi_1(z) - i\phi_2(z))}} =$$

$$= \frac{1}{\overline{g(z)}} = I(g(z)).$$

On the other hand, if f and g satisfy (a) and (b), then it is easy to show that

$$\phi(z) = \frac{1}{z^2} \overline{\phi(I(z))} .$$

It follows that, if $w = I(z) = -1/\bar{z}^2$,

$$(1.21)\qquad \alpha_z = \phi(z)dz = \overline{\phi(w)}d\bar{w} = \overline{\alpha_w}\,.$$

Hence,

$$\tilde{x}(I(p)) = \mathrm{Re}\int_{p_o}^{I(p)} \alpha_z dz = \mathrm{Re}\int_{p_o}^{I(p_o)} \alpha_z dz + \mathrm{Re}\int_{I(p_o)}^{I(p)} \alpha_z dz =$$

$$= v_o + \mathrm{Re}\int_{p_o}^{p} \overline{\alpha_w} = v_o + \tilde{x}(p).$$

Now, $\tilde{x}(p) = \tilde{x}(I(I(p))) = v_o + \tilde{x}(I(p)) = 2v_o + \tilde{x}(p)$. Therefore, $v_o = 0$ and $\tilde{x}(I(p)) = \tilde{x}(p)$, thus completing the proof of the lemma.

Let us now return to the proof of the theorem. We know that $z = 0$, $z = \infty$, $z = a$ and $z = -1/a$ correspond to the ends of $\tilde{M}$. We may rotate $S^2(1)$ in such way that $g(0) = 0$. It follows from the previous lemma that $g(\infty) = \infty$ and that $g(-1/a) = -1/\overline{g(a)}$. Furthermore, our choices were such that the branching order of g at $z=0$ is greater than or equal to the branching order of g at $z=a$. We are going to make use of the forms

$$(1.22)\qquad \beta_1 = \frac{1}{2} f g^2 dz, \qquad \beta_2 = \frac{1}{2} f dz \quad \text{and} \quad \beta_3 = fg\, dz\,.$$

These are related to the form $\alpha = (\alpha_1, \alpha_2, \alpha_3) = \phi(z)dz$ considered above, by

$$(1.23)\qquad \alpha_1 = \beta_2 - \beta_1, \qquad \alpha_2 = i(\beta_2 + \beta_1) \quad \text{and} \quad \alpha_3 = \beta_3\,.$$

In the sequel we will need the following

(1.24) LEMMA. *If* β_1 *or* β_2 *has a zero residue at* $z = 0$, *then both* β_1 *and* β_2 *are exact. If* β_3 *has a zero residue at* $z = 0$, *then* β_3 *is exact.*

<u>Proof of the lemma:</u> Assume that $\mathrm{Res}(\beta_1)_{z=0} = 0$. Then,

$$\mathrm{Res}(\alpha_1)_{z=0} = \mathrm{Res}(\alpha_2)_{z=0} = \mathrm{Res}(-i\alpha_2)_{z=0} .$$

Since α_1 and α_2 do not have real residues at $z = 0$, then

$$\mathrm{Res}(\alpha_1)_{z=0} = \mathrm{Res}(\alpha_2)_{z=0} = \mathrm{Res}(\beta_2)_{z=0} = 0 . \tag{1.25}$$

It is now a consequence of (1.21) above that $\mathrm{Res}(\alpha_1)_{z=\infty} = \mathrm{Res}(\alpha_2)_{z=\infty} = 0$. From (1.23), we then conclude that $\mathrm{Res}(\beta_1)_{z=\infty} = \mathrm{Res}(\beta_2)_{z=\infty} = 0$.

Set $A_j = \mathrm{Res}(\alpha_j)_{z=a}$, $j = 1,2$. We know that each A_j is purely imaginary. Let γ be a closed curve that is the boundary of a domain D containing $z = 0$ and $z = a$, positively oriented as the boundary of $M-D$ and invariant by I. By using (1.25) we obtain:

$$A_j = \int_\gamma \alpha_j = \int_{I(\gamma)} \alpha_j = \int_\gamma \overline{\alpha_j} = \overline{\int_\gamma \alpha_j} = \overline{A_j} , \tag{1.26}$$

where we have made use of (1.21) for the third equality. Hence, A_j is real. Since A_j is purely imaginary,

$$A_j = 0 . \tag{1.27}$$

It is now a consequence of (1.21) that

$$\mathrm{Res}(\alpha_j)_{z=I(a)} = 0 . \tag{1.28}$$

Hence α_1 and α_2 are exact and, consequently, β_1 and β_2 are also exact. An analogous and even simpler argument can be applied to the case of β_3, thus completing the proof of the lemma.

Returning to the proof of the theorem we are going to consider the various alternatives for the functions f and g.

<u>Case 1</u>: g <u>has a zero or order</u> 3 <u>at</u> $z = 0$.

Since the total curvature of x is -6π, then the total curvature of $\tilde{x}$ is -12π, and so g covers $S^2(1)$ three times. Hence, $g(z) = az^3$. Since $g(I(z)) = I(g(z))$, then $|a| = 1$. After a rotation of $\tilde{x}(\tilde{M})$ around the z-axis of R^3, we may assume that

$$g(z) = z^3.$$

For M to be complete, f must have poles at 0, a and $-1/a$ of order at least two (according to III 2.4) and must satisfy (b) of Lemma (1.20) above. Thus,

$$f(z) = b/z^2(z-a)^2(z+\tfrac{1}{a})^2 ,$$

where b is some purely imaginary constant. For this choice, observe that β_1 has residue zero at $z = 0$. Then, by Lemma (1.24) β_1 and β_2 are exact. On the other hand, a direct computation shows that $\text{Res}(\beta_2)_{z=0} = 2b(a-1/a)$. Therefore, $a = 1$. But then, $\text{Res}(\beta_2)_{z=1} = 3b/4$. This is a contradiction with the fact that β_2 is exact.

<u>Case 2</u>: g <u>has a zero of order</u> 2 <u>at</u> $z = 0$.

Then, $g(z) = z^2h(z)$. Since g covers each point of $S^2(1)$ three times, then h has exactly one zero and one pole and, by condition (a) of (1.20),

$$g(z) = cz^2(z+1/\bar{b})(z-b),$$

where $c = |b|$. If $a = b$, then

$$f(z) = d/z^2(z+1/a)^2,$$

where d is purely imaginary. Since $\text{Res}(\beta_1)_{z=0} = 0$, then β_1 and β_2 are exact. However, $\text{Res}(\beta_2)_{z=0} = -2a^3d = 0$, which is impossible.

Hence $a \neq b$ and we obtain

$$f(z) = d(z-b)^2/z^2(z-a)^2(z+1/a)^2 .$$

Again, we have $\mathrm{Res}(\beta_1)_{z=0} = 0$ and, consequently, β_1 and β_2 are exact. Applying Lemma (1.15) to $\beta_2 = \frac{1}{2} f(z)dz$, with $d_i = 0$, we obtain $a \neq 1$ and $b = a/(1-a^2)$. The same lemma applied to β_2, with $d_i = a$, yields

$$a-b = \frac{a^3 + a}{2a^2 + 1} .$$

Using the value of b already determined we have

$$a - \frac{a}{1-a^2} = \frac{a^3+a}{2a^2+1} .$$

This equation is equivalent to $a^4 + a^2 + 1 = 0$, which has no real solution. Thus, we have arrived again to a contradiction.

<u>Case 3</u>: g <u>has a zero of order one at</u> $z = 0$

A similar analysis to the one in Case 2 guarantees that

$$g(z) = \frac{cz(z+1/\bar{b})(z+1/\bar{t})}{(z-b)(z-t)} ,$$

where $c = |bt|$. If $a = b$, then

$$f(z) = d(z-t)^2/z^2(z+1/a)^2 .$$

Since $\mathrm{Res}(\beta_1)_{z=0} = 0$, the forms β_1 and β_2 are exact. Applying Lemma (1.15) to β_2 with $d_i = 0$ yields $t = -1/a$ and, hence, g has degree one, instead of degree three. This contradiction guarantees that $a \neq b$. In a similar way we conclude that $a \neq t$. Thus,

$$f(z) = \frac{d(z-b)^2(z-t)^2}{z^2(z-a)^2(z+1/a)^2} ,$$

where

$$\bar{d} = -db^2t^2/|bt|^2 . \tag{1.29}$$

Since $\operatorname{Res}(\beta_1)_{z=0} = 0$, β_1 and β_2 are exact. Application of the Lemma (1.15) to β_2 with d_i equal to 0, a and $-1/a$, yields

$$\frac{1}{b} + \frac{1}{t} = \frac{1-a^2}{a} ,$$

$$\frac{a-b}{|a-b|^2} + \frac{a-t}{|a-t|^2} = \frac{2a^2+1}{a^3+a} , \tag{1.30}$$

$$\frac{ab+1}{|ab+1|^2} + \frac{at+1}{|at+1|^2} = \frac{a^2+2}{a^2+1} .$$

If $b = b_1+ib_2$ and $t = t_1+it_2$, then, a study of the imaginary part of the equations (1.30) yields

$$\frac{b_2}{b_1^2 + b_2^2} = \frac{-t_2}{t_1^2 + t_2^2} ,$$

$$\frac{b_2}{(a-b_1)^2 + b_2^2} = \frac{-t_2}{(a-t_1)^2 + t_2^2} , \tag{1.31}$$

$$\frac{ab_2}{(1+ab_1)^2 + a^2b_2^2} = \frac{-at_2}{(1+at_1)^2 + a^2t_2^2} .$$

It follows that

$$\frac{a - 2b_1}{b_2} = - \frac{a - 2t_1}{t_2} ,$$

(1.32)

$$\frac{1 + 2ab_1}{ab_2} = - \frac{1 + 2at_1}{at_2} .$$

Thus,

$$(a-2b_1)(1+2at_1) = (a-2t_1)(1+2ab_1),$$

from which we obtain

$$a^2(t_1-b_1) = b_1-t_1.$$

Therefore, $t_1 = b_1$. Substitution of this equality in (1.32) above give us $b_2 = -t_2$, and so

$$t = \bar{b}. \tag{1.33}$$

It follows from (1.29) that $\bar{d} = -d$ and so d is purely imaginary. On the other hand, since $\beta_3 = fgdz$ has no real periods, and $\mathrm{Res}(\beta_3)_{z=0} = cd$, then cd must be real. Since c is real, then d is real. Thus $d = 0$, what is not possible.

Since all possibilities for g are covered by the cases 1, 2 and 3, and in each case we have arrived to a contradiction, we have proved the theorem.

It is clear that Lemmas (1.20) and (1.24) above can be applied to any immersion from the Möbius strip into R^3. For example, they can be used to prove the following

(1.34) PROPOSITION. There exist a unique complete minimal immersion of the Möbius strip into R^3 with total curvature -6π.

By unique we mean: unique up to reparametrization of the Möbius strip and up to rigid motion of R^3. One such example was exhibited in (IV 2). Set $M = \mathbb{P}^2 - \{p_0\}$ and let $x: M \to R^3$ represent another complete minimal immersion of the Möbius strip with total curvature -6π. Let $\pi: \tilde{M} = S^2(1) - \{q,-q\} \to M$ be the oriented two-sheeted covering of M. The mapping $\tilde{x} = x \circ \pi$ is also minimal and complete. Let $f, g\colon S^2(1) \to S^2(1)$ represent the functions associated to $\tilde{x}$ by the Weierstrass representation. From Lemma (1.20) we have that $g(I(z)) = I(g(z))$. By a rotation of $\tilde{x}(\tilde{M})$ in R^3 we may assume that g has a zero at one of the ends of $\tilde{M}$ and a pole at the other one. Changing the coordinates of $\tilde{M}$, we may also assume that $g(0) = 0$ and $g(\infty) = \infty$. Since the total curvature of x is -6π, g covers $S^2(1)$ three times. Let us now examine the various possibilities for g.

Case 1: g has one zero of order 3 at $z = 0$.

In this case $g(z) = az^3$. Since $g(I(z)) = I(g(z))$, $|a| = 1$. After a rotation of $\tilde{x}(\tilde{M})$ around the z-axis in R^3, we may assume that

$$g(z) = z^3 .$$

Since $\tilde{M}$ is complete, f must have a pole at $z = 0$, whose order must be at least two (according to (III 2.4)) and must also satisfy (b) of Lemma (1.20). Thus,

$$f(z) = d/z^4 ,$$

where d is purely imaginary. It follows that $\alpha_3 = fg\,dz$ has a nonzero real period around $z = 0$. Therefore, this case can not occur.

Case 2: g has one zero of order two at $z = 0$.

In this case $g(z) = z^2 h(z)$, with $h(0) \neq 0$. Since g covers each point of $S^2(1)$ three times, then h must have exactly one zero and one pole. Furthermore, by (a) of (1.20), we have

$$g(z) = cz^2(z+1/\bar{a})/(z-a) ,$$

where $c = |a| \neq 0$. By a rotation of $\tilde{x}(\tilde{M})$ around the z-axis in R^3, we may assume that a is a positive real number. Since $\tilde{M}$ is complete, f must have a pole at $z = 0$. Furthermore, f must have a zero of order two at $z = a$, and must satisfy (b) of Lemma (1.20). Then,

$$f(z) = d(z-a)^2/z^4 ,$$

where d is purely imaginary. A simple computation shows that $\mathrm{Res}(\alpha_3)_{z=0} = d(1-a^2)$. Since α_3 has no real periods, then $a = 1$. The functions f and g above give rise to the example already exhibited in (IV 2).

Case 3: g has one zero of order one at $z = 0$.

In this case we obtain

$$g(z) = cz \frac{(z+1/\bar{a})(z+1/\bar{b})}{(z-a)(z-b)} ,$$

where $|c| = |ab|$. The function f must have zeroes of order two at $z = a$ and $z = b$ and a pole at $z = 0$. From Lemma (1.20), it follows that

$$f(z) = d(z-a)^2(z-b)^2/z^4 ,$$

where $\bar{d} = -dc^2/\bar{a}^2\bar{b}^2$. Observe that, since $\tilde{M}$ has only two ends, then (1.26) can be applied with $\gamma(t) = e^{it}$, $0 \leq t \leq 2$, to obtain $\mathrm{Res}(\beta_j)_{z=0} = 0$, $1 \leq j \leq 3$. A direct computation of the residues give

$$\mathrm{Res}(\beta_1)_{z=0} = c^2 d(\bar{a}+\bar{b})/\bar{a}^2\bar{b}^2 = - \bar{d}(\bar{a}+\bar{b}),$$

$$\mathrm{Res}(\beta_2)_{z=0} = -d(a+b),$$

$$\mathrm{Res}(\beta_3)_{z=0} = dc(ab + \frac{1}{\bar{a}\bar{b}} - (a+b) + (\frac{1}{\bar{a}} + \frac{1}{\bar{b}})).$$

It follows that $a+b = 0$ and $dc(ab+1/\bar{a}\bar{b}) = 0$. This last equality holds only if $d = 0$ or $c = 0$, that can not occur.

Thus, the proposition is proved.

REFERENCES

L.V. Ahlfors and L. Sario

1. Riemann Surfaces, Princeton Univ. Press, 1960.

J.L. Barbosa and M.P. do Carmo

1. Helicoids, catenoids, and minimal hypersurfaces of R^n invariant by an P-parameter group of motions, An. Acad. Brasil. de Ciências 53 (1981), 403-408.

J.L. Barbosa, M. Dajczer and L.P.M. Jorge

1. Minimal ruled submanifold in spaces of constant curvature, Ind. Univ. Math. J., 33 (1984), 531-547.

S. Bernstein

1. Sur les surfaces définies au moyen de leur courbure moyenne ou totale, Ann. Sci. l'École Norm. Sup., 27 (1910), 233-256.
2. Sur les équations du calcul des variations, Ann. Sci. l'École Norm. Sup. (3) 29 (1912), 431-485.
3. Sur un théorème de Géométrie et ses applications aux équations aux dérivées partielles du type elliptique, Comm. de la Soc. Math. de Kharkov (2-ème sér.) 15 (1915-1917), 38-45.

L. Bianchi

1. Lezioni di geometria differenziale, Spoerri, Pisa, 1894.

D.E. Blair and J.R. Vanstone

1. A generalization of the helicoid, in "Minimal Submanifolds and Geodesics", (pp.13-16), Kaigai Publications, Tokyo 1978.

E. Bombiere, E. de Giorgi and E. Giusti

1. Minimal cones and the Bernstein problem, Inventiones Math. 7 (1969), 243-268.

E. Calabi

1. Quelques applications l'analyse complexe aux surfaces d'aire minima (together with Topics in Complex Manifolds by Hugo Rossi) Les Presses de l'Université de Montréal, 1968.

M.P. do Carmo

1. Differential Geometry of Curves and Surfaces, Prentice-Hall, Inc., 1976.

M.P. do Carmo and M. Dajczer

1. Rotation hypersurfaces in spaces of constant curvature, Trans. Amer. Math. Soc., 277 (1983), 685-709.

M.P. do Carmo and C.K. Peng

1. Stable complete minimal surfaces in R^3 are planes, Bull. Amer. Math. Soc., 1 (1979), 903-906.

E. Catalan

1. Journal de Mathém. 7 (1842), 203.

C.C. Chen and P.A.Q. Simões

1. Superfícies Mínimas do R^n, Escola de Geometria Diferencial, Campinas, 1980.

C.C. Chen and F. Gackstatter

1. Elliptic and hyperelliptic functions and complete minimal surfaces with handles, IME-USP, nº 27, 1981.

C.C. Chen

1. A characterization of the Catenoid, An. Acad. Brasil. de Ciências, 51 (1979), 1-3.
2. Elliptic functions and non-existence of complete minimal surfaces of certain type, Proc. of A.M.S. vol. 79, nº 2 (1980), 289-293.
3. Total curvature and topological structure of complete minimal surfaces, IME-USP, nº 8, 1980.

S.S. Chern and R. Osserman

1. Complete minimal surfaces in Euclidean n-space, J. d'Analyse Math., 19 (1967), 15-34.

S.S. Chern

1. An elementary proof of the existence of isothermal parameters on a surface, Proc. Amer. Math. Soc. 6 (1955), 771-782.
2. Minimal surfaces in a Euclidean space of N dimensions, Differential and Combinatorial Topology, a Symposium in Honor of Marston Morse, Princeton Univ. Press (1965), 187-198.

C.J. Costa

1. Imersões mínimas completas em R^3 do gênero um e curvatura total finita, Ph.D. thesis, IMPA, 1982.
2. Example of a complete minimal immersion in R^3 of genus one and three embedded ends, Bol. Soc. Bras. Mat., 15 (1984), 47-54.

G. Darboux

1. Leçons sur la théorie générale des surfaces, vol. I, 3rd. ed., Chelsea, N.Y., 1972.

A. Enneper

1. Zeitschrift für Mathm. u. Physik, 9 (1864), 108.

L.P. Eisenhart

1. A treatise on the differential geometry of curves and surfaces, Dover publications, 1909.
2. An introduction to Differential Geometry, Princeton University Press, 1947.

H.M. Farkas and I. Kra

1. Riemann Surfaces, Springer-Verlag, 1980.

J.R. Feitosa

1. Imersões Mínimas Completas no R^3 de Gênero um, com Curvatura Total Finita, MS Thesis, UFC, 1984.

F. Gackstätter

1. Über die Dimension einer Minimalfläche und zur Ungleichung von St. Cohn-Vossen, Arch. Rat. Mech. Anal. 61 (1976), 141-152.
2. Über Abelsche Minimalflächen, Math. Nachr. 74 (1976), 157-165.

Henneberg

1. Annali di Matem. 9 (1878), 54-57

D. Hoffmann and W.H. Meeks III

1. A complete minimal surface in R^3 with genus one and three ends, preprint.
2. Complete embedded minimal surfaces of finite total curvature, preprint.

A. Huber

1. On subharmonic functions and differential geometry in the large, Comm. Math. Helv. 32 (1957), 13-72.

L.P.M. Jorge and F. Xavier

1. A complete minimal surface in R^3 between two parallel planes, Ann. of Math., 102 (1980), 204-206.
2. On the existence of a complete bounded minimal surface in R^n, Bol. Soc. Bras. Mat., vol. 10 (1979), 171-183.

L.P.M. Jorge and W.H. Meeks, III

1. The topology of complete minimal surfaces of finite total Gaussian Curvature, Topology, 22 (1983), 203-221.

T. Klotz and L. Sario

1. Existence of Complete Minimal Surfaces of arbitrary connectivity and genus, Proc. N.A.S., 54 (1965), 42-44.

J.L. Lagrange

1. Oeuvres, vol. I, (1760), 335.

H.B. Lawson, Jr.

1. Lectures on Minimal Submanifolds, vol. 1, Publish or Perish, Inc., 1980.

W.H. Meeks, III

1. The geometry and conformal structure of triply periodic minimal surfaces in R^3, Ph.D. Thesis, Univ. of California, Berkeley, 1975.
2. Lectures on Plateau's Problem, Escola de Geometria Diferencial, Fortaleza, 1978.
3. The topological uniqueness of minimal surfaces in three dimensional Euclidean space
4. The classification of complete minimal surfaces with total curvature greater than -8π, Duke Math. Journal, 48, 3 (1981), 523-535.

J.B. Meusnier

1. Mémoire sur la courbure des surfaces, Mémoires des savans étrangers 10 (lu 1776), 1785, 477-510.

H. Mori

1. Minimal surfaces of revolution in H^3 and their stability properties, Ind. Math. J. 30 (1981), 787-794.

E.H. Neville

1. Elliptic Functions, A Primer Pergamon Press, 1971.

J.C.C. Nitsche

1. On new results in the theory of minimal surfaces, Bull. Amer. Math. Soc., 71 (1965), 195-270.
2. Vorlesungen über Minimalflächen, Springer Verlag, 1975.

Maria Elisa G.G. de Oliveira

1. Superfícies Mínimas não-orientáveis no R^n, Ph.D. Thesis, IMEUSP, 1984.

H. Omori

1. Isometric Immersions of Riemannian Manifolds, J. Math. Soc. Japan, 19 (1967), 205-214.

R. Osserman

1. A survey of Minimal Surfaces, Van Nostrand, 1969.
2. Minimal Surfaces in the large, Comm. Math. Helv., 35 (1961), 65-76.
3. On Complete Minimal Surfaces, Arch. Rational Mech. and Anal. 13 (1963), 392-404.
4. Proof of a conjecture of Niremberg, Comm. Pure Appl. Math. 12 (1959), 229-232.
5. Global Properties of minimal surfaces in E^3 and E^n, Ann. of Math. 80 (1964), 340-364.
6. Global Properties of classical Minimal Surfaces, Duke Math. J., 32 (1965), 565-574.

T. Radó

1. On the Problem of Plateau, Springer, Berlin, 1933.

W. Rudin

1. Real and Complex Analysis, Tata McGraw-Hill, New Delhi, 1966.

P. Scherk

1. Crelle's Journal f. Mathem. 13 (1835).

R.N. Schoen

1. Uniqueness, Symmetry, and Embeddedness of Minimal Surfaces, preprint.

H.A. Schwartz

1. Ges. Mathematische Abhandlungen I, Verlag Julius Springer, 1890.

C.L. Siegel

1. Topics in Complex Function Theory, vol. I, Willey-Interscience, 1969.

M. Spivak

1. A comprehensive introduction to Differential Geometry, Publish or Perish, Inc., Berkeley, 1979.

G. Springer

1. Introduction to Riemann Surfaces, Addison-Wesley Publishing Co., 1957.

D.J. Struik

1. Lectures on classical Differential Geometry, Addison-Wesley, 1950.

K. Voss

1. Über vollständige Minimalflächen, l'Enseignement Math. 10 (1964), 316-317.

K. Weierstrass

1. Monatsber. Berlin Akad., 1866.

F. Xavier

1. The Gauss map of a complete, non-flat minimal surface cannot omit 7 points of the sphere, Annals of Math., 113 (1981), 211-214.

S.T. Yau

1. Some function-theoretic properties of complete Riemannian manifolds and their applications to Geometry, Indiana Univ. Math. J., 25 (1976), 659-670.

INDEX